VOLUME ONE HUNDRED AND EIGHTY EIGHT

PROGRESS IN MOLECULAR BIOLOGY AND TRANSLATIONAL SCIENCE

Molecular Biology and Clinical Medicine in the Age of Politicization

VOLUME ONE HUNDRED AND EIGHTY EIGHT

PROGRESS IN MOLECULAR BIOLOGY AND TRANSLATIONAL SCIENCE

Molecular Biology and Clinical Medicine in the Age of Politicization

Edited by

TOBY BOLSEN
Department of Political Science,
Georgia State University, Atlanta, GA, United States

RISA PALM
Urban Studies Institute,
Georgia State University, Atlanta, GA, United States

ELSEVIER

ACADEMIC PRESS
An imprint of Elsevier

Academic Press is an imprint of Elsevier
50 Hampshire Street, 5th Floor, Cambridge, MA 02139, United States
525 B Street, Suite 1650, San Diego, CA 92101, United States
The Boulevard, Langford Lane, Kidlington, Oxford OX5 1GB, United Kingdom
125 London Wall, London EC2Y 5AS, United Kingdom

First edition 2022

Notices
Knowledge and best practice in this field are constantly changing. As new research and experience broaden our understanding, changes in research methods, professional practices, or medical treatment may become necessary.

Practitioners and researchers must always rely on their own experience and knowledge in evaluating and using any information, methods, compounds, or experiments described herein. In using such information or methods they should be mindful of their own safety and the safety of others, including parties for whom they have a professional responsibility.

To the fullest extent of the law, neither the Publisher nor the authors, contributors, or editors, assume any liability for any injury and/or damage to persons or property as a matter of products liability, negligence or otherwise, or from any use or operation of any methods, products, instructions, or ideas contained in the material herein.

ISBN: 978-0-323-99439-2
ISSN: 1877-1173

For information on all Academic Press publications
visit our website at https://www.elsevier.com/books-and-journals

Publisher: Zoe Kruze
Acquisitions Editor: Ashlie M. Jackman
Developmental Editor: Jhon Michael Peñano
Production Project Manager: James Selvam
Cover Designer: Matthew Limbert

Typeset by STRAIVE, India

Contents

Contributors ix
Preface xi

1. Science, politics and institutional design in regulation **1**
Thomas O. McGarity and Wendy E. Wagner

1. Institutional design for science-policy in the US 4
2. Science and politics in the federal government's reaction to the COVID-19 pandemic 8
3. Institutional design in practice: Multiple pathways to politicizing science 14
4. A better way 17
5. Conclusion 24

References 24

2. Beyond the sheltering academic silo: Norms for scientists' participation in policy **29**
Karen L. Akerlof

1. Introduction 29
2. Protective silos: Science and objectivity 31
3. The evolving relationship between science and policy 32
4. Norms for engaging in hyperpolitical settings: Legislatures 35
5. Interviews with congressional staff regarding appropriate scientist roles in policy 36
6. Re-imagining scientists' engagement with policy 40
7. Conclusion 41

References 41

3. Effects of politicization on the practice of science **45**
Hannah Schmid-Petri, Nils Bienzeisler, and Arista Beseler

1. Introduction 46
2. Conceptualization and definition of politicization of science 47
3. The politicization of science in the triangular relationship of politics, media, and science and its consequences in the practice of science 49
4. Conclusion and practical implications: What can be done to address politicization? 57

References 59

4. Politics v. science: How President Trump's war on science impacted public health and environmental regulation **65**

Romany M. Webb and Lauren Kurtz

1. The Silencing Science Tracker 66
2. Anti-science actions under Trump 67
3. The Biden administration's approach to science 74
4. Where to from here? Restoring the role of science 76

References 77

5. Politicization and COVID-19 vaccine resistance in the U.S. **81**

Toby Bolsen and Risa Palm

1. The development of vaccines for COVID-19 82
2. How did the COVID-19 virus itself become politicized? 83
3. Politicization of the COVID-19 vaccine 85
4. Emergence of variants and continued politicization 89
5. What does social science research conclude about promoting vaccine acceptance 90
6. What does politicization mean for practitioners of science? 92

References 94

6. Effects of politicized media coverage: Experimental evidence from the HPV vaccine and COVID-19 **101**

Erika Franklin Fowler, Rebekah H. Nagler, Darshana Banka, and Sarah E. Gollust

1. The science of guidelines and updating health recommendations 103
2. Concerns over politicization, dimensions of the concept, and strategies for mitigation 104
3. Study objectives 107
4. Data and methods 109
5. Results 115
6. Discussion 125

Appendix 128

References 131

7. Vax attacks: How conspiracy theory belief undermines vaccine support **135**

Christina E. Farhart, Ella Douglas-Durham, Krissy Lunz Trujillo, and Joseph A. Vitriol

1. Introduction 136
2. Literature review and framework 137

3. Hypotheses 143
4. Data and methods 143
5. Results 148
6. Discussion and conclusion 155
Appendix A. Question Wording 157
References 164

8. Communicating CRISPR: Challenges and opportunities in engaging the public **171**
Elizabeth S. Baik, Abraham Koshy, and Bruce W. Hardy

1. Introduction 172
2. The challenge from the speed of advancement 173
3. What CRISPR means to different publics: The challenge of lack of quality data 174
4. Engaging diverse and underrepresented publics: Shared knowledge production and decision-making 178
5. Public engagement of CRISPR in the age of politicization 181
6. Ways to overcome identity protective cognition 183
7. Conclusion 187
References 188

9. Moral conviction: A challenge in the age of science politicization **195**
Robin Bayes

1. What is moral conviction? 196
2. Why does politicization matter for moral convictions? 198
3. How prominent are moral convictions in public opinion on science issues? 202
4. How might strategic communication create moral conviction? 205
5. Moral conviction: A challenge to the foundations of science? 208
References 210

10. Strategic science communication in the age of politicization **215**
Todd P. Newman

1. Introduction 215
2. Scientists attitudes and behavior toward public communication 217
3. Strategic science communication: Connecting goals, objectives, and tactics 221
4. The brand of science and why it matters 223
5. Conclusion 227
Acknowledgments 228
References 228

Index *231*

Contributors

Karen L. Akerlof
Department of Environmental Science & Policy, George Mason University, Fairfax, VA, United States

Elizabeth S. Baik
Klein College of Media and Communication, Temple University, Philadelphia, PA, United States

Darshana Banka
Wesleyan University, Middletown, CT, United States

Robin Bayes
Department of Political Science, Northwestern University, Evanston, IL, United States

Arista Beseler
University of Passau, Passau, Germany

Nils Bienzeisler
Karlsruhe Institute of Technology (KIT), Karlsruhe, Germany

Toby Bolsen
Department of Political Science, Georgia State University, Atlanta, GA, United States

Ella Douglas-Durham
Harvard T.H. Chan School of Public Health, Department of Social and Behavioral Sciences, Cambridge, MA, United States

Christina E. Farhart
Carleton College, Department of Political Science, Northfield, MN, United States

Erika Franklin Fowler
Wesleyan University, Middletown, CT, United States

Sarah E. Gollust
University of Minnesota, Minneapolis, MN, United States

Bruce W. Hardy
Klein College of Media and Communication, Temple University, Philadelphia, PA, United States

Abraham Koshy
Klein College of Media and Communication, Temple University, Philadelphia, PA, United States

Lauren Kurtz
Climate Science Legal Defense Fund, New York, NY, United States

Krissy Lunz Trujillo
Harvard Kennedy School, Shorenstein Center, Cambridge; Northeastern University, Network Science Institute, Boston, MA, United States

Thomas O. McGarity
Joe R. and Teresa Lozano Long Professor in Administrative Law and Richard Dale Endowed Chair, University of Texas School of Law, Austin, TX, United States

Rebekah H. Nagler
University of Minnesota, Minneapolis, MN, United States

Todd P. Newman
Department of Life Sciences Communication, University of Wisconsin-Madison, Madison, WI, United States

Risa Palm
Urban Studies Institute, Georgia State University, Atlanta, GA, United States

Hannah Schmid-Petri
University of Passau, Passau, Germany

Joseph A. Vitriol
Stony Brook University, Department of Political Science, Stony Brook, NY, United States

Wendy E. Wagner
Joe R. and Teresa Lozano Long Professor in Administrative Law and Richard Dale Endowed Chair, University of Texas School of Law, Austin, TX, United States

Romany M. Webb
Sabin Center for Climate Change Law, New York, NY, United States

Preface

In the latter half of the 19th century, as reported in the then-new journal *Nature*, science was being lauded as the most significant achievement of society. Benjamin Disraeli stated that the "revolutions of science ... have had more effect than any political causes, which have changed the position and prospects of mankind more than all the conquests and all the codes and all the legislators that ever lived."[1] Similarly, in 1876, Louis Pasteur praised science as knowing no country "because knowledge belongs to humanity and is the torch which illuminates the world."[2] Yet, 150 years later, science confronts a public crisis of trust.[3,4] As Feine and Jakubovics recently summarized it, "the pandemic has highlighted a widespread mistrust for scientific advice in populations across the world. In the United States, recent antiscience political agendas have appeared to increase this lack of trust, sometimes to the point of open hostility."[5 (p.4)]

Suhay has noted that acceptance of scientific knowledge depends on "trust in the scientific enterprise," and that such trust varies widely in populations, much of which is attributable to political identity.[6 (p.1)] At the same time, the lack of trust in science and scientists results in doubts about scientific findings, increasing the difficulty of communicating information about issues such as climate change, genetically modified foods, stem cell research, CRISPR technology, vaccinations, nuclear waste disposal, and so forth.[7–11]

The results of this increasing lack of trust and politicization of science itself result in clear challenges for molecular biology and clinical medicine. First, politics and public interest groups can influence the scientific agenda. Second, research funding can be distorted by political or economic goals, slowing progress in otherwise promising lines of research. Third, politicization of science can lead to pressure on scientists themselves to censor their own findings or outright rejection of scientific findings and research methods.

There are several examples, both positive and negative, of lobbying by public- and special-interest groups affecting both basic science and clinical research. On the positive side, Suhay noted the impact of citizen groups on attention to both muscular dystrophy and AIDS.[6,12] On the other hand, Legg et al. documented the impact of multinational corporations and industry interest groups to shape the research agenda.[13] Medical research is also influenced by the vast lobbying efforts of the pharmaceutical industry that can exert pressure on elected officials and the FDA to fast-track drugs or approve questionable drugs.[14]

The second issue is the fact that political and economic goals can influence funding available for research. On the issue of climate change research, for example, Farrell found empirical evidence of "the actual social arrangements within which large-scale scientific (mis)information is generated, and the important role private funding plays in shaping the actual ideological content of scientific information that is written and amplified."[15 (p.93)] Research funding and directions in this sense can be shaped by the kinds of requests for proposals developed as well as evaluation panels. This is exemplified in the efforts of the sugar industry to gain access "to the expert panel of the US National Institute of Dental Research as a means of shaping the US government's research priorities towards interventions that did not involve restricting the consumption of sugar [and] purposefully create misinformation, doubt, and ignorance (or "agnogenesis")—to, inter alia, obscure the harms of industry products and practices, and oppose environmental, occupational and public health regulation that could threaten corporate profits."[13 (p.2),16–20] As Suhay summarizes it, "by only funding scientific research that was likely to counter arguments for regulation (research casting doubt on the smoking-cancer link in the former case and on anthropogenic climate change in the latter), these industries and their ideological allies successfully tilted the pool of knowledge in their favor."[6 (p.20)]

Finally, outside interests can put pressure on scientists themselves not to publish their results or to edit them in ways that do not offend government or industry sponsors. One mode of censoring scientists is the use of misconduct charges aimed not to maintain scientific integrity but instead to intimidate the scientists themselves: In these cases, the charges are often made up and not intended to maintain scientific integrity. Misconduct charges typically initiate a formalized process, which can burden the researcher and threaten their credibility.[21–24] Another mode is the creation of rules for data sharing and transparency that are unrealistic in scope and would restrict the timely sharing of data and findings.[25] These techniques, and others, can create doubt about the independence and validity of the science, resulting in further loss of trust by the public in science, public health, and its practitioners.[26]

Why are politics and science intertwined?

Druckman described the "crisis of politicization within and beyond science" as the result of three factors: first, that science is easily politicized because it is inherently uncertain and "questioning extant knowledge is part of the research process"; second, that human reasoning, and particularly the tendency to "seek out information that confirms prior beliefs" or motivated reasoning, results in people "dismissing evidence inconsistent with prior opinions, regardless of objective accuracy"; and third, that technological evolution, permitting not only the greater ease of both data collection and analysis but also the profusion of misinformation from "expert" pundits on the Internet, has led to greater challenges to scientific findings, further exacerbating the communication of scientific uncertainty.[27 (p.615)] Bolsen and Druckman have defined how the intertwining of science and politics becomes a "communicative tactic" as the inherent uncertainty of science is emphasized to cast doubt on scientific consensus.[28] Wise adds that politicization not only involves the insertion of political perspectives in federal regulatory and research agencies but also "continuing commercialization of academic research threatens to make universities and federal research institutions look similarly political."[29 (p.5)]

The purpose and organization of this volume

The subject of this volume, politicization of science, is different from those typically covered in Progress in Molecular Biology and Translational Science. It is so because of recent unprecedented attacks on the veracity and the value of science, not only in the United States but also throughout the world. These attacks threaten not only public support for, and funding of, science but also its very foundations. Molecular biologists, physician-scientists, clinicians, stem cell researchers, and other researchers cannot afford to ignore the profound impacts that nonscientific factors have on the conduct of their research. Science is affected by the society, the economy, and the political system within which it operates. These factors affect both what science focuses on and how science is conducted. If scientists and clinicians are to continue to provide the knowledge needed to develop desperately needed vaccines (e.g., to fight COVID-19); cures for cancer, heart disease, breast cancer, and stroke; medical interventions tailored to individual patients (precision medicine); and myriad other revolutionary advancements, they must understand and seek to counteract politicization of science. We seek here to provide to readers of Progress in Molecular Biology and Translational Science a deeper understanding of nonscientific forces impinging on their work, which we hope will provide them with the tools necessary to combat attempts to destroy science.

The 10 chapters in this volume have been divided into three sections. The first provides a general introduction to the notion of science politicization and its realization in recent attempts to shape or limit scientific research. The second section provides a series of specific illustrations of politicization in vaccine acceptance. The third section provides strategies by which scientists can overcome this politicization to promote the regaining of trust in science and in scientists by the general public.

Section 1: Politicization of science

Chapter 1: Science, politics and institutional design in regulation

McGarity and Wagner examine the design of the US federal bureaucracy, with a focus on politicization of science in the US Executive Branch, to shed light on when and how politicization occurs in environmental and health protection. The chapter makes clear that scientific decision-making processes are often subject to political interference and control by appointees. It introduces the reader to the architecture of science-based decision-making in the US federal bureaucracy, reviews specific incidents of politicization that have occurred in the past and in the recent response to the COVID-19 pandemic, and highlights possibilities for remodeling institutions in ways that prevent and penalize the politicization of science. A key takeaway is that it is essential to increase the autonomy and independence of agency scientists from stakeholder and political control.

Chapter 2: Beyond the sheltering academic silo: Norms for scientists' participation in policy

Akerlof describes how the role of science vis-à-vis society has changed over decades in dramatic ways that inherently make it more political, but that scientists may not necessarily realize or acknowledge. The chapter focuses specifically on legislatures as sites for science advice to decision-makers. It reports the results from interviews with legislative staff in the US Congress who serve as gatekeepers for the flow of policy-relevant information. Members of Congress embrace a wide range of norms for scientists' involvement, but they also cite concerns about the effects of redefining these boundaries on scientists' authority.

Chapter 3: Effects of politicization on the practice of science

Schmid-Petri, Bienzeisler, and Beseler sharpen the concept of science's politicization by explaining how it is a "communicative act" by which science comes to acquire political meaning in democratic countries. The chapter reviews how political actors, the media, and scientists can drive politicization on any matter where science is relevant to a political debate. It offers guidance for best practices among scientists who seek to balance communicating their important findings and advocating for specific political policies they might support.

Chapter 4: Politics v. science: How President Trump's war on science impacted public health and environmental regulation

Webb and Kurtz review federal, state, and local "anti-science actions" recorded from November 8, 2016, to January 20, 2021, by the *Silencing Science Tracker*—a joint project of the Climate Science Legal Defense Fund and Columbia's Sabin Center for Climate Change. Federal actions that contributed to a "war on science" under President Trump's administration included government- and self-censorship, budget cuts, personnel changes, research hindrance, and misrepresentation of scientific studies. The chapter describes how various forms of scientific research were portrayed as inherently political under the Trump administration. It highlights the pressing need for reforms to better protect federal scientists from political interference and to identify ways to restore public trust in science.

Section 2: Politicization and the acceptance of vaccines

Chapter 5: Politicization and COVID-19 vaccine resistance in the U.S.

Bolsen and Palm describe the roots of politicization surrounding COVID-19 vaccines and factors that have contributed to resistance and hesitancy over time in the United States. They show that variability in the statements of political elites and media coverage in response to the outbreak of the disease—and over time during the development, approval, and rollout of FDA-approved vaccines—contributed to polarization and vaccine hesitancy and resistance among segments of the public. The chapter addresses the emergence of COVID-19 variants and ongoing polarization, as well as possible avenues for overcoming politicization to increase the uptake of life-saving vaccines.

Chapter 6: Effects of politicized media coverage: Experimental evidence from the HPV vaccine and COVID-19

Drawing on an experimental dataset embedded in a large pilot survey conducted in the United States in the summer of 2020, Fowler, Nagler, Banka, and Gollust find that exposure to different "dimensions of politicization" increases people's negative emotional responses, and confusion, toward health-related information about the HPV vaccine and COVID-19 vaccines. In addition, they find that an intervention designed to defend against various forms of politicization did not blunt the confusion it generated toward the vaccines. They also report some evidence of "spillover effects" whereby exposure to political controversy surrounding the HPV vaccine led to decreased support for immunization programs more broadly. The chapter clarifies different meanings of politicization and the challenges it poses to the uptake of life-saving vaccines.

Chapter 7: Vax attacks: How conspiracy theory belief undermines vaccine support

Farhart, Douglas-Durham, Lunz Trujillo, and Vitriol evaluated the factors that influence vaccine hesitancy in two surveys conducted in the United States in fall 2020. They demonstrate that general conspiratorial thinking, endorsement of COVID-19 conspiracy theories, antiintellectualism, and partisan-/ideological-identity are associated with reduced support for the uptake of COVID-19 vaccines. The findings accentuate the importance of research that identifies ways to overcome these barriers to public health.

Section 3: Effective communication of science

Chapter 8: Communicating CRISPR: Challenges and opportunities in engaging the public

Baik, Koshy, and Hardy describe the rapid advance of CRISPR gene-editing technologies and review what is known about public opinion regarding the use of these technologies. They find a major gap in research on the public's understanding of these technologies, as well as limited media coverage. Community-engagement-based participatory research is recommended as a framework for decision-making that engages the public. This entails ongoing dialogue with community members that recognizes the complex and value-laden choices that are involved with the use of CRISPR technologies. The authors offer recommendations for best practices for facilitating public engagement, which will be especially important as CRISPR technologies continue to advance and potentially become politicized due to the potential applications of these technologies.

Chapter 9: Moral conviction: A challenge in the age of science politicization

Bayes explores why moral conviction matters for understanding public response to scientific information. The chapter presents the results from a pilot survey conducted in September 2021 designed to evaluate the role moral convictions play in shaping people's attitudes across several issues related to biomedicine. It discusses how people come to hold moral convictions on these issues and challenges that this can pose for science practitioners.

Chapter 10: Strategic science communication in the age of politicization

Newman provides an overview of the research on how scientists think about communication and public engagement, and how the public thinks about science and scientists. The chapter examines the extent to which and in what way scientists across various disciplines engage in strategic communication practices in their own communication and public engagement, as well as provides an overview on how scientists and the public think and feel about science. Through unpacking these different connections to science, several recommendations for scientists' communication and engagement within the current context of "postnormal" science are provided.

Toby Bolsen
Department of Political Science, Georgia State University, Atlanta, GA 30302, United States
Risa Palm
Urban Studies Institute, Georgia State University, Atlanta, GA 30302, United States

References

1. Disraeli B. Banquet speech, Glasgow. *Nature*. 1873;9:71.
2. Pasteur L. *Banquet Toast at the International Congress of Sericulture, Milan, Italy, as Translated in René Dubos, Louis Pasteur, Free Lance of Science*; 1960:85. Banquet date identified in Maurice B. Strauss, Familiar Medical Quotations (1968):519.

3. Otto SL. *The War on Science: Who's Waging It, Why It Matters, What We Can Do About It.* Minneapolis, MN: Milkweed Editions; 2016.
4. Parikh S. Why we must rebuild trust in science. *Trend Mag.* 2021. https://www.pewtrusts.org/en/trend/archive/winter-2021/why-we-must-rebuild-trust-in-science.
5. Feine J, Jakubovics N. Science in the spotlight: a crisis of confidence? *JDR Clin Trans Res.* 2021;6(1):4–7.
6. Suhay E. The politics of scientific knowledge. In: *Oxford Research Encyclopedia of Communication*; 2017. https://doi.org/10.1093/acrefore/9780190228613.013.107.
7. Suhay E, Druckman JN. The politics of science: political values and the production, communication, and reception of scientific knowledge. *Ann Am Acad Pol Soc Sci.* 2015;658(1):6–15. https://doi.org/10.1177/0002716214559004.
8. Douglas HE. Untangling values, ideologies, and reasons. *Ann Am Acad Pol Soc Sci.* 2015;658(March):296–306.
9. Nisbet E, Cooper KE, Garrett RK. The partisan brain: how dissonant science messages lead conservatives and liberals to (dis)trust science. *Ann Am Acad Pol Soc Sci.* 2015;658 (March):36–66.
10. Kahan DM. Why we are poles apart on climate change. *Nature.* 2012;488(7411):255.
11. Lupia A. Communicating science in politicized environments. *Proc Natl Acad Sci USA.* 2013;110(suppl 3):14048–14054.
12. Bucchi M, Neresini F. Science and public participation. In: *The Handbook of Science and Technology Studies.* vol 3; 2008:449–472.
13. Legg T, Hatchard J, Gilmore AB. The science for profit model—how and why corporations influence science and the use of science in policy and practice. *PLoS One.* 2021;16(6):e0253272.
14. Unsal O. Two faces of corporate lobbying: evidence from the pharmaceutical industry. *North Am J Econ Finance.* 2020;51:100858.
15. Farrell J. Corporate funding and ideological polarization about climate change. *Proc Natl Acad Sci USA.* 2016;113(1):92–97.
16. McGarity TO, Wagner WE. *Bending Science: How Special Interests Corrupt Public Health Research.* Cambridge, MA: Harvard University Press; 2008.
17. Mialon M, Julia C, Hercberg S. The policy dystopia model adapted to the food industry: the example of the Nutri-Score saga in France. *World Nutr J.* 2018;9:109–120.
18. Proctor RN, Schiebinger L. *Agnotology. The Making and Unmaking of Ignorance.* Stanford, CA: Stanford University Press; 2008.
19. Savell E, Fooks G, Gilmore AB. How does the alcohol industry attempt to influence marketing regulations? A systematic review. *Addiction.* 2015;111:18–32. https://doi.org/10.1111/add.13048. 26173765 (web archive link).
20. Savell E, Gilmore AB, Fooks G. How does the tobacco industry attempt to influence marketing regulations? A systematic review. *PLoS One.* 2014;9(2):e87389. https://doi.org/10.1371/journal.pone.0087389. 24505286 (web archive link).
21. Kuehn RR. Suppression of environmental science. *Am J Law Med.* 2004;30(2 & 3):333–369.
22. Lewandowsky S, Bishop D. Don't let transparency damage science. *Nature.* 2016;529 (7587):459–461.
23. Lewandowsky S, Mann ME, Brown NJ, Friedman H. Science and the public: debate, denial, and scepticism. *J Soc Political Psych.* 2016;4(2):537–553.
24. Väliverronen E, Saikkonen S. Freedom of expression challenged: scientists' perspectives on hidden forms of suppression and self-censorship. *Sci Technol Hum Values.* 2021;46 (6):1172–1200.
25. Goobie GC, Rice MB, Carlsten C. The EPA's 'strengthening transparency in pivotal science' rule: don't let history repeat itself. *Ann Am Thor Soc.* 2021.

26. Goldberg RF, Vandenberg LN. The science of spin: targeted strategies to manufacture doubt with detrimental effects on environmental and public health. *J Environ Health*. 2021;20(1):1–11.
27. Druckman JN. The crisis of politicization within and beyond science. *Nat Hum Behav*. 2017;1(9):615–617.
28. Bolsen T, Druckman JN. Counteracting the politicization of science. *J Commun*. 2015;65(5):745–769.
29. Wise MN. Thoughts on politicization of science through commercialization. In: *Science in the Context of Application*. Dordrecht: Springer; 2011:283–299.

CHAPTER ONE

Science, politics and institutional design in regulation

Thomas O. McGarity and Wendy E. Wagner*
Joe R. and Teresa Lozano Long Professor in Administrative Law and Richard Dale Endowed Chair, University of Texas School of Law, Austin, TX, United States
*Corresponding author: e-mail address: wwagner@law.utexas.edu

Contents

1. Institutional design for science-policy in the US 4
   1.1 Political officials can often control agency scientific analyses 5
   1.2 Staffing and budget 7
   1.3 All together 8
2. Science and politics in the federal government's reaction to the COVID-19 pandemic 8
   2.1 Downplaying risks 9
   2.2 Early testing for COVID-19 9
   2.3 Antibody tests 10
   2.4 Chloroquine and hydroxychloroquine 11
   2.5 Premature reopening 13
3. Institutional design in practice: Multiple pathways to politicizing science 14
   3.1 Manipulating science in pursuit of an agenda 14
   3.2 Manipulating agency resources and organization 15
   3.3 Manipulating science advisory boards 15
   3.4 Well-endowed stakeholders monopolize public deliberations 16
4. A better way 17
   4.1 An ideal institutional design for science-policy 18
   4.2 A working example of a reformed design 19
5. Conclusion 24
References 24

Abstract

The politicization of science is a recurring phenomenon in US federal policymaking that is explained in part by the unstructured, collaborative nature of decision-making in most science-intensive US regulatory programs. In this chapter we spotlight some of the most significant worries arising from this longstanding approach to U.S. institutional design for expert agencies and offer recommendations for stronger institutional structures to help guard against the politicization of science.

Progress in Molecular Biology and Translational Science, Volume 188
ISSN 1877-1173
https://doi.org/10.1016/bs.pmbts.2021.11.002

For political officials who must make unpopular decisions regarding the protection of public health or the environment, scientific information is a mixed blessing. On the one hand, scientific research provides invaluable information on how alternative policies might play out in the real world, which in turn allows an official to carefully craft an effective response. A rigorous scientific analysis also provides a political official with some cover, allowing her to shrug off controversy by observing that "the science made me do it."

On the other hand, that same scientific analysis, if taken seriously, may significantly constrain choices and leave only very unpleasant ones as plausible options. Scientific research may reveal, for example, that the air in certain industrial corridors is unsafe for residential life, but the cost of reducing pollution to acceptable levels will force some companies out of business. Or it may reveal that climate change can only be slowed by restricting carbon emissions in ways that powerful corporations and other campaign contributors vehemently oppose. Conceding that science should inform policymaking also forces a political official to abdicate some of her power to scientists; the political official is no longer in complete control of the framing or analyses.

Politicization of science—which we define in this chapter as either mischaracterizing research findings or dismissing rigorous research as unscientific in order to advance a predetermined political end—provides an escape hatch for a political official who feels boxed in by the scientific evidence. By taking charge of what the science says, political officials can have their cake and eat it too. When the science is in line with a preferred policy, that research provides a testament to the wisdom of the political official's choices. When science gets in the way, it can simply be massaged, altered, or dismissed. In this way, as long as a political official reserves a hand in controlling what the science says, science becomes a kind of "good fairy" to support her decisions.

Given the risks that science might be distorted or compromised by policymakers to justify a preferred outcome, we might expect US government institutions to be designed in ways that limit political control over the agencies' scientific analyses. We might imagine, for example, that agency scientists would enjoy independence to at least state what the scientific literature says on basic issues of health protection that involve life and death consequences. We might also expect that agency procedures for issuing rules or approving products would provide protection against political interference with the underlying scientific assessments.

However, as we detail in this chapter, the decision processes in the US often lack these basic protections of scientific independence. Rather, most agency processes in the US are designed in accordance with the "good fairy" model. Specifically, these processes allow political officials within the agency—and in some cases even the President himself—to generally exert unilateral control on all types of work the agency produces, including analyses of the relevant scientific evidence.

As long as institutions are designed in ways that do not insulate experts' scientific assessments or fact-check key political statements against an established scientific record, the good fairy model can become irresistible to some political officials. They can control inconvenient science by politicizing it. And because the issues are technical and complex, they can often accomplish this politicization without getting caught in the act.

This chapter examines the design of the US federal bureaucracy to better understand when and how the politicization of science occurs in the area of environmental and health protection. Thankfully, in most policy decisions, an agency's scientific analyses will not be politicized because the analyses do not impose constraints on political officials that are significant enough to warrant this type of mischief. In a few other cases, a political official may even note that his or her decision is at odds with a staff's robust scientific analysis, but insist openly that other permissible considerations such as economics over-ride public health concerns or the environment. However, at least some of the time, a political official will find that the easiest course is to exercise control over the science itself.

Our examination of the politicization of science in the US Executive Branch unfolds in four parts. After a brief orientation to the institutional architecture in the US for science-intensive decision-making, we review some of the politicization of science incidents that have surfaced over the last few decades, starting with the most visible recent efforts epitomized by President Trump's COVID response. We then trace a number of less overt strategies that the White House and political officials have used to politicize science over time. The chapter then closes on a hopeful note with a proposal for reform. While science and politics cannot be strictly separated, there are ways to design institutional processes to create some barrier between foundational scientific analyses and difficult political choices. Indeed, we discuss one agency process that in fact does introduce firewalls to protect staff scientists' analyses against the politicization of the science upon which they are based.

Based on our analysis, we believe that that as long as science can be controlled by decision makers, it will be. Many policy choices are complex and laden with uncertainty, and there is often no clear or easy way to resolve health and environmental controversies. Ethical norms and good behavior on behalf of political officials thus remain a dubious means of curbing the politicization of science problem in the United States. One essential step to minimize these risks, is to remodel our institutional design in ways that prevent and penalize the politicization of science.

1. Institutional design for science-policy in the US

In the United States, the politicization of science by political officials is but one manifestation of a much larger problem. Indeed, efforts by industry and other affected parties to manipulate science in ends-oriented ways are legendary.[1–4] A veritable playbook of tactics for controlling the science that informs policy has emerged over decades of environmental and health policymaking.[4,5]

In response, considerable effort has been dedicated to devising ways to protect applied scientific research and analyses from external influence. Science journals, for their part, have devised ways to protect the integrity of research from external influences by, for example, requiring scientists to disclose conflicts of interest to provide some assurance that they approached their research in a skeptical, open-minded way so as to allow readers to judge for themselves whether conflicts may have affected their conclusions.[6–8] The importance of scientific "independence" in policymaking is also underscored in academic and related work on the design of expert bureaucracies.[9] In a series of reports, the Organization for Economic Cooperation and Development (OECD) identifies independence from political and stakeholder control to be *the* most important institutional design feature in advancing agency expertise.[10,11] The importance of agency autonomy has also been spotlighted by bipartisan task forces as a key design principle for administrative processes.[12,13]

Each of these efforts to preserve the integrity of science underscore the import of scientists' autonomy in conducting separate, foundational scientific analyses to inform policy. Without a public record of the staff scientists' independent scientific findings, there is no limit to the ability of political officials to engage in invisible, ends-oriented tinkering in the scientific record and to promulgate policies that have been carefully crafted to obscure where the science left off and the policymaking began.[4,14] Unscrupulous political actors can even start with a desired end goal and work back to subtly manipulate each phase of the scientific analysis to support a desired conclusions.

However, the institutional processes that govern agency expertise in the United States move in precisely the opposite direction from these best practice ideals. Indeed, some key institutional structures appear antithetical to allowing, much less encouraging, true expert advice within our agencies.

1.1 Political officials can often control agency scientific analyses

In most federal agencies in the United States, there is a clear line of authority that positions one or more politically appointed officials at the apex of the entire agency; this includes authority over all of the agencies' scientific analyses.[15] In this way, political officials are not only responsible for the agency policies and rules that are informed by scientific analyses, but they can also establish internal procedures for whether or how staff scientists can publish their research, speak to the press, attend meetings, etc.

On the science-policy issues that arise in regulation, the appropriateness of making political officials (rather than staff scientists), responsible for the final agency decisions is beyond question. It has been long accepted that applied scientific analyses that inform policies are rife with value questions and that these value judgments are not isolated at the end of a science-intensive decision but occur throughout.[16] Even simple risk assessments, for example predicting the risks arising from exposure to a carcinogen, require multiple "policy" judgments about the veracity and relevance of individual studies; the populations of greatest concern; how to extrapolate from individual studies or even from test animals to a larger system, and so forth.[16,17] The resulting multiple layers of judgment mean that the synthesis of the underlying scientific literature, as well as designing the studies in the first place, involve not only the run-of-the-mill consensual judgments from within the scientific community, but also policy choices about the nature of alternative assumptions, which in turn will vary in risk aversion and protectiveness.

Accordingly, it is also well-accepted that agency scientists should not operate with complete independence in their work because aspects of their technical analyses may involve value judgments. In determining the weights to be afforded to different studies in a research synthesis, for example, there are underlying policy judgments associated with how to weight studies that find effects against those that do not; how to weight studies produced with conflicts of interest against those with no conflicts; and how to weight studies based on methods that may generate different types of risks of false positives. For that reason, agency processes often direct that scientific staff analyses not be released publicly until the analyses have been vetted internally; it is important that the agency speak with one voice.[18,19] Likewise, rogue agency personnel should be prevented from making statements that would mislead or confuse the public.

When implemented without any internal limits on the political official's power over the scientific analyses, however, this institutional design may allow for the politicization of science in ways that can be difficult, if not impossible, to detect. Political appointees retain the prerogative to intervene and may insist on alterations to the technical analyses that are unwarranted by the science.[20,21]

Unfortunately, in many agency decision-making processes, the scientific and policy staff routinely work together to develop technical analyses and recommendations, without any constraints in place to protect the integrity of the scientific advice.[21] Indeed, in some programs, the very first science-intensive report that the agency makes public is jointly offered between the scientific staff and political officials.[21] This structure for the decision-process ultimately gives political officials the decisive vote, while giving the appearance of a decision that is grounded in rigorous scientific research.

Under this institutional design, agency scientific staff are in a position of subservience to policymakers. If technical analyses present unwelcome findings, political officials are free to exercise their executive discretion and demand adjustments. And when they do this, there is generally no way for those outside the agency—short of staff whistleblowers— to assess the integrity and independence of the resulting technical analyses. The Freedom of Information Act (FOIA) in fact protects these collaborative discussions from public disclosure as "deliberative process."[22,23] Outsiders are thus unable to assess the role that politics might have played in an analysis that is presented as primarily technical.[21]

The problems with this design have not gone unnoticed. Particularly during the George W. Bush Presidency, accounts of politicization of science inside agencies, such as the censorship of prominent agency scientists, led to a rash of protests and concerns by the scientific community. In response, President Obama early his Administration directed all agencies to develop scientific integrity policies to ensure some modicum of independence to scientific staff in conducting basic research and to protect them from harassment.[24,25] But most of the agency science integrity policies resulting from Obama's Order are unenforceable and in any event fall short of correcting the larger institutional structure that allows for political control over characterization of the staff's findings.[25] Even in agencies with the most robust scientific integrity policies, staff may be able to publish findings in their own name, but in cases where there is an internal disagreement over what the evidence suggests for policy or how best to characterize the scientists' findings, the political officials retain significant control.[25]

1.2 Staffing and budget

Agency expertise is only as good as the quality of the scientific staff the agency hires and the resources available to do the work. Under the existing design of decision processes, however, political appointees have the discretion to make these personnel and budget decisions in a political, ends-oriented fashion.[25,26] While there are justifications for some political control over agency staffing, the implications for ensuring a rigorous, independent group of scientists within an expert agency are, at the very least, compromised and only serve to create more opportunity for political officials to control the science. As Gillian Metzger observes, "massively underfunding the EPA without altering its statutory responsibilities or repealing environmental rules ··· [even] begin to look constitutionally suspect" from the proposition that "delegated power must be faithfully executed."[27]

Under existing institutional arrangements, this political control over the staffing of agency experts in the Executive Branch does not stop with the agency's own fulltime employees, but can extend to the creation and composition of the science advisory boards assembled to peer review the agency's technical analysis.[14,28] While federal law mandates that the scientists selected to serve on these scientific advisory boards must be "balanced," there is also a soft presumption that since they are operating as peer reviewers, those serving on the boards will also be respected, mainstream scientists who represent the views of the diverse scientific community.[13,29,30] In practice, however, there is no legal prohibition against stacking science advisory boards in ends-oriented ways with scientists with clear ideological views who are not necessarily respected within the scientific community.[28] The fact that these boards are funded and operate at the discretion of a political administrator also means that they can be discontinued if the risks of peer review of the agency's science-intensive findings are too threatening to their political goals.[25]

1.3 All together

In view of this permissive institutional architecture for science-intensive decision-making, there are at least three primary ways that science can be politicized. First, some and perhaps most of the political manipulation of science may occur out of sight as officials place pressure on or alter the features of staff scientific analyses that get in the way of a desired political goal. Second, political officials may exert control over the hiring, management, and budget of scientific staff. This signals to staffers that, if they do not fall in line and collaborate with political officials, they may come to regret it. Third, a political official can even establish new rules for how science is used by an agency and in so doing bias outcomes or discourage respected scientists from serving in the agency. For example, an administrator might revise procedures in ways that will ensure that peer review is done only by hand-picked "experts" who will support desired results. By changing the ground rules for scientific analyses, bias can be introduced into all the expert work the staff does, not just the work on an individual project.

2. Science and politics in the federal government's reaction to the COVID-19 pandemic

In this section we survey more blatant examples of the politicization of science in recent years. Bold attempts by President Trump to mischaracterize and ignore expert advice during the COVID crisis were not subtly slipped into policy decisions in ways that went unnoticed. Indeed, President Trump's politicization efforts often resulted in media circuses and political sideshows.[14,31]

Nevertheless, President Trump's unrelenting efforts to politicize science in ends-oriented ways can be traced to pre-existing problems with our existing institutional design, which allows political control over agency expertise. Seen through this lens, President Trump's politicization of science was enabled in part by a pre-existing institutional design that permits this behavior. In fact, we will see that prior Presidents have exploited this permissive design as well, albeit generally in more subtle ways. Thus, while Trump's politicization of science was more out in the open and obvious—and perhaps even more deadly—than those of his predecessors, it is a recurring phenomenon in the US Executive Branch.

Throughout the Trump Administration's response to the COVID-19 pandemic, scientific evidence was highly relevant to all phases of decision-making, but the relevant government officials on many occasions either downplayed the significance of that information or ignored it altogether in their determination to advance President Trump's volatile policy prescriptions. Perhaps the defining characteristic of the Trump administration's approach to dealing with the coronavirus crisis was its unwillingness to base its decisions on the best available science.[32] Rather than convene expert panels or insist on scientific peer review for recommended policies, the experts were subservient to the Chief Executive and his non-scientific assessment of the evidence.

2.1 Downplaying risks

Worried that fear of a pandemic would slow down a bustling economy upon which he was basing his re-election message, President Trump downplayed the risks posed by the coronavirus. In late January, Trump announced that the COVID-19 outbreak was "totally under control."[33] On February 2, he assured the public that his partial ban on travel from China "pretty much shut it down coming from China."[33] In mid-February, he proclaimed that warmer weather in April would kill the virus, a position that was contradicted by experts at the World Health Organization.[33] On February 27, he predicted that "one day—it's like a miracle—it will disappear."[34]

Unfortunately, the president's reassurances were inconsistent with the scientific data and sound scientific analysis. When a CDC scientist in late February told the media that COVID-19 was in fact spreading rapidly throughout the country and the real question was "how many people in this country will have this severe illness,"[33] the resulting dip in the stock market infuriated President Trump and inspired Health and Human Services Secretary Alex Azar to prohibit CDC employees from publicly commenting on the pandemic without explicit authorization.[34]

2.2 Early testing for COVID-19

One way to stop the spread of the disease early in outbreak was to test people showing symptoms, quarantining those who tested positive, and testing people who came into close contact with people who contracted the disease. High level politically appointed officials in the Centers for Disease Control (CDC) decided that a test that its scientists had developed would be the only permissible test and that samples could only be sent to a few designated laboratories.[35] Scientists at the Food and Drug Administration (FDA) wanted to work with scientists in universities and commercial laboratories to get more tests approved by that agency, but the Secretary of Health and Human Services, a political appointee, refused to allow that initiative to go forward out of fear that it would yield so many positive results that the message would be inconsistent with the president's ongoing assurances that the administration had the COVID-19 pandemic under control.[35]

2.3 Antibody tests

After suffering intense criticism for the approach that it took toward diagnostic tests for determining whether a subject had contracted COVID-19, the FDA took the opposite approach toward approving tests for whether the subject had produced antibodies, a possible indication that the subject had developed an immunity to the disease.[36] Once again, it declined to follow the science. Many viewed such tests as the key to opening up the economy because people who tested positive for the antibodies would presumably be immune to the disease and able to work.[36] President Trump was therefore anxious to get as many antibody tests as possible on the market because he wanted to open up the economy as quickly as possible.[37] Many experts, however, believed that antibody tests were not necessarily the key to safely reopening the economy because there was no proof that the presence of antibodies in the blood guaranteed immunity. A false negative would just mean that the subject would still have to observe the necessary precautions even though they may have been unnecessary. A false positive, however, could give the subject a false sense of security that the coronavirus no longer posed a risk.[36]

On March 16, the FDA issued guidance allowing test developers to market antibody tests without having received an emergency use authorization so long as they self-validated the tests, notified the FDA, and indicated on the label that the test had not been approved by the FDA.[36] The announcement resulted in a proliferation of more than 160 unreviewed, but aggressively marketed tests.[38] Only 13 tests had received official emergency use authorizations (EUAs) from FDA.[38] Some of the unapproved tests had been developed by university laboratories and were of high quality.[36] Others were of questionable accuracy, deeply flawed, or even fraudulent. One test that the City of Laredo, Texas purchased in large quantities turned out to be only 20% accurate, though it boasted a 93–97% accuracy rate.[37] Some came from China, India and other countries with few assurances of reliability.[36] And some made unsubstantiated claims that persons testing positive were immune and could discontinue social distancing.[36] Now experts argued that

FDA had allowed too many unapproved tests on the market too rapidly. The result was confusion, wasted resources expended on worthless tests, and a false sense of security among many individuals who were not in fact immune.[37] Things got so bad that state attorneys general began to file lawsuits against fraudulent manufacturers.[39] In this instance, the FDA sacrificed quality for speed to advance the Trump agenda, and the market became "a wild, wild West."[40]

The FDA reacted to media reports on the proliferation of inaccurate tests and a critical letter from the chairman of the House subcommittee on Economic and Consumer Policy with a May 5 announcement that the agency would require all antibody tests to receive full-fledged EUAs.[38] And in order to receive an EUA, the company would have to demonstrate that the test was 90% accurate in detecting antibodies and 95% accurate in concluding that antibodies were not present.[38] If companies filed an application for an EUA within 10 days their products could remain on the market until the FDA acted on the applications. Products from companies failing to meet the deadline and products failing to receive EUAs would be subject to seizure.[40] This action did not clear the market of faulty and fraudulent products, but it did begin the process of ensuring that coronavirus tests were reasonably accurate.[39] How effective it would be at that task would depend on how effectively the FDA enforced the new requirement.

At the same time, the FDA stepped up its approval of EUAs for products that met its requirements. In early May, it authorized a test developed by Roche Holding that was 98% accurate and could be produced in huge quantities at Roche's 100 laboratories throughout the country.[41] Another pharmaceutical company, Abbot Labs received authorization for a test that it could produce in quantities of 60 million units per month by June.[42]

2.4 Chloroquine and hydroxychloroquine

When the Trump administration policymakers finally accepted in mid-March 2020 the existence of the pandemic that the scientists had been warning of since mid-January, the search was on for a magic bullet that would allow the fading economy to recover. Since vaccines would not be available for at least 8 months under the rosiest of scenarios, the solution would have to lie in finding an effective treatment for those who contracted the disease. Acting on a tip from his friend Larry Ellision, the CEO of the database vendor Oracle, President Trump seized on the anti-malarial drugs chloroquine and hydroxychloroquine as the cure.[43] An assistant secretary at the Department of Health and Human Services (HHS) directed Dr. Rick Bright, the head of the department's Biomedical Advanced Research and Development Authority (BARDA), to "take a look" at chloroquine.[44] The BARDA scientists were skeptical, noting that no clinical studies had been conducted to test the efficacy of either chloroquine or hydroxychloroquine in treating COVID-19.[44] Health and Human Services Secretary Alex Azar, however, failed to mention the scientists' concerns when he attended a meeting with the president and FDA Commissioner Stephen Hahn, CDC Director Robert Redfield and Dr. Anthony Fauci to discuss the drugs.[43] The president wanted the FDA to approve and promote the drugs to provide a "shot of inspiration" to the public.[45] But Drs. Hahn, Redfield and Fauci urged the president to hold off promoting the drugs until they had been shown to be safe and effective in randomized clinical trials.[43]

Trump ignored the scientists. The next day he held a press conference to announce that hydroxychloroquine could be a "game changer" in the nation's efforts to treat COVID-19 patients.[46] He erroneously proclaimed that the FDA had been "so great" in immediately approving the drug to treat the disease.[47] Doctors could in fact prescribe chloroquine and hydroxychloroquine for the unapproved use of treating COVID-19,[43] but the FDA had not approved them for that use.[48] The President's announcement precipitated a massive run on the drugs that made them difficult to find for the people suffering from the diseases for which they were approved.[49]

In late March, the White House ordered Secretary Azar to make hydroxychloroquine available to the public through the Strategic National Stockpile.[44] Azar in turn directed Dr. Bright, the head of BARDA, to work with the FDA to develop an investigational new drug protocol that would make the drug available to patients of doctors who participated in the program through a phone app that Oracle would develop.[44] The BARDA scientists, however, were concerned by studies indicating that hydroxychloroquine could cause irregular heart rhythms and even death in some people. They worried about the safety of a program disguised as a scientific study for making the drugs available to patients through doctors outside of closely supervised hospital settings.[44]

Dr. Bright and Dr. Janet Woodcock, the head of the FDA's drug approval office, came up with an intermediate solution under which the FDA would grant an Emergency Use Authorization that would make the drugs available for use in hospitals to treat patients diagnosed with COVID-19, a far more limited use than the investigational new drug option.[44]

This subtle, but important difference in legal status had no effect on the President's attempts to promote the drug as a cure that could "clean out the lungs."[50] The White House instructed Dr. Bright's superiors at HHS to "flood NY and NJ" with hydroxychloroquine.[44] When the deputy director in charge of the strategic national stockpile responded that the FDA's emergency use authorization limited the drugs to hospitalized patients, a high-level politically appointed official ordered him do it anyway.[44] The stockpile then sent millions of doses of hydroxychloroquine to hospitals for use by patients being treated for COVID-19, but they also went to pharmacies for use by outpatients in direct violation of the emergency use authorization.[51]

It soon became clear that the drugs were not the miracle cure that the President had hoped for. Reports from Wuhan, China indicated that hydroxychloroquine was not effective in treating COVID-19,[52] and doctors in France halted a clinical trial of hydroxychloroquine when electrocardiogram monitoring indicated that it was causing an irregular heartbeat in one of the patients.[53] This precipitated a warning from the FDA in late April that chloroquine and hydroxychloroquine could cause abnormal heart rhythms and a dangerously high pulse rate in patients taking the drugs to treat COVID-19.[54] After two major studies published in the *Journal of the American Medical Association* and the *New England Journal of Medicine* found no evidence that the drugs lowered the risk of intubation or death,[55] the FDA in mid-June withdrew the emergency use authorization.[56] Fortunately, most doctors had already quit prescribing the drugs for COVID-19.[57] And the President, who had gotten far out ahead of the scientific evidence, stopped touting the drugs.[58]

2.5 Premature reopening

Against the advice of his scientific advisors, President Trump on April 29 declared that the country had "crossed a big boundary and much better days are ahead."[59] States began to loosen restrictions, and the president resumed large campaign rallies where masks were offered but not required.[60,61] As the second wave of infections kicked in and the infection rate increased rapidly to 60,000 per day, the White House decided to look to different experts for advice. Drs. Fauci and Birx were sidelined as the White House sought out Fox News commentator Dr. Scott Atlas, a professor of radiology at Stanford University. He advised policymakers that they should shelter the elderly and immune-compromised in place and allow the disease to spread through the rest of the population until herd immunity kicked in.[35] Wearing masks, in his view, was futile.[62] The second wave peaked in July, and the infection rate declined gradually until September when it rose rapidly in a much larger third wave that peaked in January when vaccinations became available.[63] Unfortunately, hospitalizations and deaths followed the same pattern, causing an unbearable strain on medical services in many states and enormous grief among those who lost friends and relatives to the disease.[64]

3. Institutional design in practice: Multiple pathways to politicizing science

President Trump's tactics for politicizing science were hardly pioneering. Over the decades, numerous political officials have taken advantage of their power to manipulate agency scientific analyses when they got in the way of preferred policies. Indeed, most of these manipulations of the scientific record surfaced from individual staff whistleblowing or anonymous reports of political meddling, which makes it likely that numerous stealth attacks on the integrity of the scientific advice likely go undetected.[18]

In this section, we lay out a variety of tactics that have emerged in decades of Executive Branch politicization of science. While some of these strategies presumably arose from the trial and error involved in controlling inconvenient scientific results, the strategies used to control science have been repeated frequently enough over time to expose recurring patterns of the subterranean politicization of science.[14,31]

3.1 Manipulating science in pursuit of an agenda

Under the United States constitution and modern administrative law, the president and his or her political appointees can control the substance of scientific input into agency decisions unaccountably. The time-worn strategies that have developed over the last 40 years share the same architectural failings; there is no scientific record against which to compare the manipulations. And, even if there were a record, there is also no expert (peer review) check on the political characterizations of those records that take liberties with the scientific facts in ways that—in the abstract—should be capable of correction.

Our review of the literature on "politicization" identified recurring strategies aimed at a form of censorship.[31] First, political officials can control and even prevent the dissemination of agency-produced science by designing the decision-making process so that the analyses and conclusions of agency scientists must receive approval from upper-level politically appointed superiors before making them public.[25] Second, they can design agency procedures to limit scientific input into agency decisions.[25,31] Third, they can manipulate the output of models the agency relies on by "tweak [ing] model inputs, assumptions, outputs, and interpretations in ways that withstand public and even scientific scrutiny, but are in fact nothing more than raw politics and outcome-oriented decisionmaking."[25] Fourth, they can subtly substitute policy for science by overruling agency experts without providing reasons for doing so, a tactic frequently employed by the White House Office of Management and Budget.[25] Finally, they can constrain the staff's analysis of the best available science by forbidding them from relying on relevant scientific studies for reasons unrelated to the science, such as the inability of researchers to make all of underlying data available to anyone who wants to see it.[25,31] All of these strategies can be shielded from public view through the agency's invocation of the deliberative process exception to FOIA. It is therefore difficult for the public to ascertain the extent to which the conclusions are based on scientific information that has been vetted by the scientific community and the extent to which they are based on policy (or political) considerations.

3.2 Manipulating agency resources and organization

Upper-level political appointees can achieve policy outcomes that do not necessarily reflect sound scientific input indirectly by exploiting the agency experts' lack control over budgets and management charts.[14] The easiest way to reduce the influence of agency staff scientists is to reduce the budgets of troublesome offices.[25] In the resulting "reduction in force," uncooperative scientists can be let go. Upper-level political appointees can also get rid of unwanted scientists by reassigning them to meaningless positions for which they are not qualified or even moving their offices to far-off places in the anticipation that many will elect to resign or retire instead of uprooting their families.[25] These strategies allow officials motivated by policy or political preferences to undermine the independence and autonomy of agency experts.

3.3 Manipulating science advisory boards

Many agencies that rely on science also rely on scientific advisory committees, composed of experts from academia, industry, and stakeholder groups, to review agency decision documents and provide advice on the scientific aspects of various regulatory actions. Some statutes, including the Clean Air Act, require agencies to seek the advice of a panel of scientists, but agencies frequently establish advisory committees on their own to improve the quality of their decisions and to provide a degree of political cover against the criticism that their decisions are not sufficiently based on science. More than 200 such scientific advisory committees assist federal agencies in bringing science to bear on issues at the intersection of science and policy.[25]

The advice that outside scientific advisory bodies provide, however, is subject to manipulation by political appointees bent on achieving outcomes that may be inconsistent with the scientific consensus. Agency decisionmakers can, of course, ignore the advice of the outside experts, though that strategy is not without political risks.[5] Members of scientific advisory committees have occasionally called out politically appointed agency decisionmakers who have tried to hide policy decisions behind the veneer of science.[25] But agency heads can accomplish their policy and political goals indirectly by changing the composition of the advisory committees by replacing scientists who are likely to provide unwanted advice with those who are likely to provide advice consistent with those goals.[14,25] During the Trump Administration, EPA Administrator Scott Pruitt attempted to permanently skew the composition of the agency's many scientific advisory committees by excluding any scientists who were currently receiving EPA research grants but allowing scientists employed by or receiving support from industry to remain on the committees, a policy that a court later overturned.[14,31,65]

3.4 Well-endowed stakeholders monopolize public deliberations

Public participation is an essential component of agency decision-making in the United States, but the process for participation established by the Administrative Procedure Act is passive. The agency publishes a notice in the Federal Register and invites interested parties to comment on the agency's proposal. The assumption is that if the opportunity to engage is available, those with the greatest information and stakes in the outcome will engage. That is, however, an unrealistic assumption in a world in which the companies subject to regulations and their trade associations have vastly more resources than the individuals who benefit from the regulations and the public interest groups who attempt to represent them.[66] The impact of the disparity in resources is especially acute in highly technical, science-intensive policymaking where asymmetric pressure can skew information, review, and litigation risks in ways that can impact the agency scientific assessments, even when agencies attempt to guard against it.

These imbalances in the opportunities for engagement have created a cottage industry of stakeholder-funded "scientific" organizations that feed off these institutional design deficiencies. Following the familiar tenets of supply and demand, a wealth of think tanks, websites, and other opportunistic private organizations can produce a continuous supply of sponsor-controlled, ends-oriented scientific information designed to wear down underfunded agency staffs.[67] Reducing the "demand" for such information by reforming the government institution will not eliminate these organizations, because they also aim to influence popular opinion in ways that turn up more diffusely in voting and partisan politics. But realigning government processes will go a long way to undercut their success in government decision-making.

4. A better way

The Trump Administration's experience with the COVID-19 pandemic is a more blatant illustration of the politicization of science that can occur in the US Executive Branch, but even in administrations that are more inclined to rely on science, we see it can be very difficult for political officials to steadfastly avoid the temptation to control science when it poses a barrier to a desired political decision. This control, as we have seen, results at least in part from unstructured institutional processes that provide few protections against the ability of policymakers to manipulate or mischaracterize the scientific record.

Rather than continue to tolerate a permeable internal boundary between scientific consensus and political preferences, we therefore propose a more impermeable firewall between science and politics on foundational issues arising in science-intensive decisions, such as what the scientific record says and whether scientific staff have the independence to report it accurately. In this restructured world, policymakers may still reach decisions that conflict with or ignore the scientific record, but they will be forced to do so in ways that openly acknowledge the conflicts or discrepancies, rather than paper them over by mischaracterizing the scientific findings.

In this final section, we introduce recommendations for a more constructive institutional approach to science-intensive policy. We then follow our general proposal with a real-world example of this preferred approach operating in practice. This example showcases the workability of our recommendation, while also flagging remaining challenges that require more attention in the future.

4.1 An ideal institutional design for science-policy

An institutional structure that will succeed in protecting science from politicization requires the creation of firewalls for key features of scientific advice, including scientific literature reviews; instituting peer review processes to "fact check" key policy pronouncements; assembling respected peer reviewers to do this work; and insulating the management of the scientific staff from political control. We discuss each feature in turn.

One fundamental and straightforward opportunity to insulate internal scientific advice from politicization arises during the scientific staff's assessment of the relevant scientific literature. Whether the agency's mandate directs it to set protective levels for chemicals, identify endangered species, or assess habitat destruction or environmental risks, the first step for staff is to summarize the existing scientific literature that bears on the policy problem.[25] A rigorous review of the scientific literature within an agency decision-process should at the very least follow the processes in place for the publication of a review article in a reputable scientific journal. The agency staff working on the review should be listed as authors with appropriate attribution, state any conflicts of interest, and be firewalled from the political officials in ways that do not allow for any influence over that fundamentally scientific assessment. The reports should be rigorously peer reviewed by respected scientists. Indeed, because the staffs' reputation will be impacted by the quality of their published report, members of the scientific staff should be eager to do high quality work and will presumably go to extra lengths to make their judgments and decisions clear and transparent. The report, interim drafts, and all peer review comments must be made public. The publication of this scientific literature review has the capacity to discipline at least some political mischaracterizations of the scientific record in subsequent steps of the policymaking exercise.

With the scientific literature review in hand, the policy decisions then begin to enter the picture. Yet even here, there is the need for some added structure to the decision process. Although a respectable scientific record should now exist, political officials can still mischaracterize it in ways that may not be detected by the public or even busy scientists. To discourage this behavior, a revised institutional approach should also introduce established scientific panels or reviewers to "fact check" policymakers' characterization of the scientific record on key issues. For example, if an official's qualitative statements about the scientific record become an important feature of a policy decision, then these statements should be peer reviewed by a respected, disinterested group of scientists. At the very least, a final policy proposal that summons the scientific literature for support should undergo scientific checking to correct any misunderstandings. Rather than provide political officials with full, unconstrained power to characterize the scientific record as they see fit, then, this added step of periodic fact checking provides a more structured approach that reigns in the freedom of politically appointed officials to politicize science.

Essential to both the first and second recommendations is the need for rigorous and trustworthy peer review.[14] Members of science advisory boards and individual peer reviewers must be highly respected scientists with expertise on the relevant issues and a commitment to disinterestedness. Political officials should play no role in selecting these scientists; the selection of reviewers should instead originate from an independent entity like the National Academies of Science, Engineering and Medicine or the Health Effects Institute.

A final reform entails insulating the hiring and management of scientific staff themselves from political control. Rather than allow political officials to control hiring, management, and budgets of scientific staff, a reformed institutional design would entrust career scientist-managers to hire and fire, make the assignments, and propose annual budgets for agency scientific departments. To ensure insulation for these career-managers, they in turn would report to an independent unit, perhaps within the Congressional Research Service or General Accounting Office.

4.2 A working example of a reformed design

A good example of at least some of these recommendations in practice is the EPA's setting of ambient standards for major air pollutants. Under the Clean Air Act, the EPA is directed to periodically revise the health-based primary national ambient air quality standards (NAAQS) and set the primary standard at a level that is "requisite to protect the public health" with an "adequate margin of safety."[68] The high visibility and cost of these standards, coupled with the extensive litigation on virtually all of the standards, led the EPA to revise its approach substantially in 2006.[21] This revised approach provides a template for the kind of idealized approach we recommend, although since judgment is still endemic to the scientists' own analysis, devising ways to explain those consensual judgments and hold them into account remains a challenge.

In its revised approach to setting the NAAQS, the EPA breaks its decision-making process into three sequential steps, each of which endeavors to isolate rigorous scientific advice from subsequent policy deliberations so that the integrity and coherence of the scientific analyses is not compromised.[21] Each step entails a separate report authored by agency technical staff who have been firewalled from the political officials (meaning there are no political controls over the analyses in these three reports). This structural separation of science and policy provides benefits not only in separating the technical assessment from the political work, but also in heightening the transparency of the expert deliberations in ways that the collaborative science-policy decision-making approach tends to obscure.[25] Each of the three staff reports is peer reviewed, often multiple times, and the agency also openly solicits comments from the public and other interested parties. The three sequential reports then establish a rigorous scientific record upon which the subsequent policy decisions can be made.[21]

However, a preliminary step in the NAAQS process—even before the scientific work begins—requires formal direction from policymakers regarding the relevant questions that scientists should resolve in their assessment of the scientific literature. These questions are further refined with public suggestions. The final, policy-framed questions then set the stage for all subsequent scientific work.[21]

With policy questions in hand, a team of scientists then undertakes a full comprehensive review of the relevant scientific literature. The work at this stage is akin to a large literature review and is called the Integrated Science Assessments or ISA.[21] Under the Clean Air Act, the ISA must "accurately reflect the latest scientific knowledge" on the health effects of the pollutant.[69] The evidence that goes into the ISAs typically comes in three primary categories: controlled human studies, epidemiological studies, and animal studies. As a result of the large body of research reviewed by the authors, the ISAs are lengthy and detailed.[21] See Fig. 1.[70] To ensure accuracy, the reports are peer reviewed, sometimes multiple times, and also reviewed by interested stakeholders to ensure that the report is not biased. The identities of both authors and peer reviewers are revealed, and they are given full credit and attribution.[21]

Even in this basic scientific analysis of the existing literature, there are still many points involving informed judgment. For example, the quality of the studies that provide the grist for evidence-based considerations can be quite

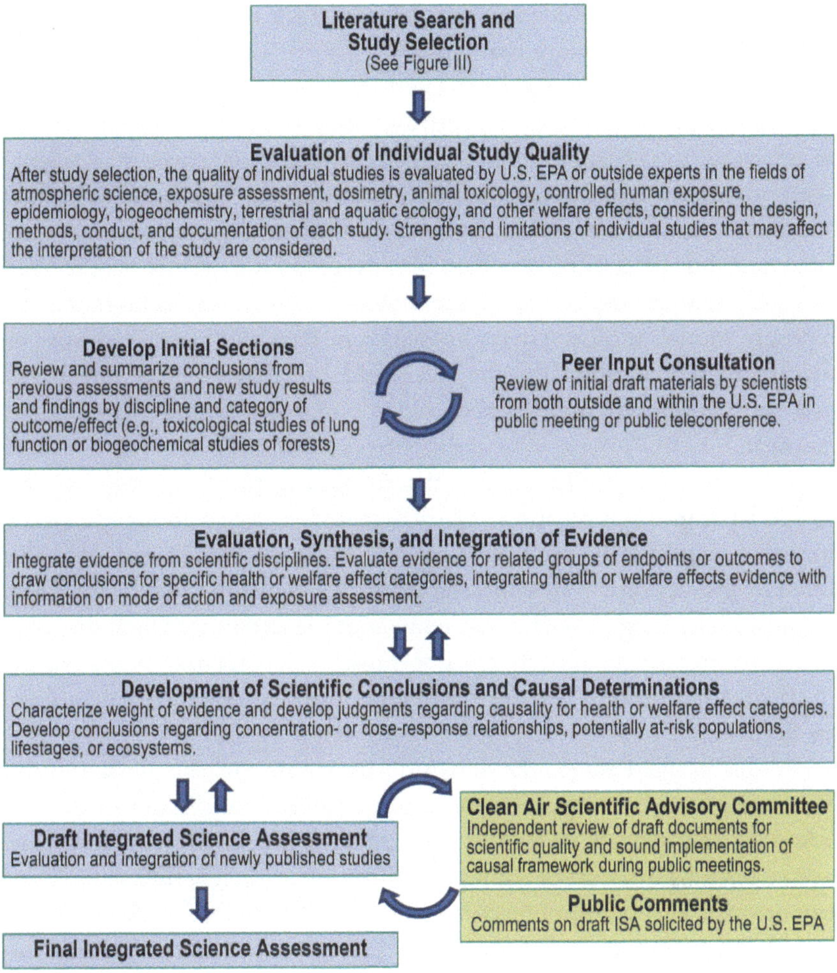

Fig. 1 Characterization of the general process for integrated science assessment development.

uneven. Some studies may be so flawed that they are entitled to no weight and are therefore not included in the ISA. Other studies may be weighted and qualified in the ISA to alert policymakers and others to their flaws, which might be considered in setting policy.[21] Fortunately, since the ISA is often subject to multiple rounds of peer review by a science advisory board, as well as public comment, many of the most significant judgments intrinsic to the experts' scientific review are scrutinized and presumably qualified. Yet the embedded, inescapable value judgments remain a concern that may warrant additional finetuning in the design of the process in the future.

A similar, science-intensive process occurs for the second report in the series—the Risk/Exposure Assessment.[21] This report involves the development of multiple quantitative risk assessments and exposure models to predict and estimate risks associated with air pollutants. Because the information on exposure to environmental pollutants is relatively sparse, staff scientists again encounter large uncertainties in assessing exposures to those pollutants.[71] Air quality monitoring devices vary in their ability to measure exposures, and thus EPA fills in the gaps in the data with exposure models, some of which are more sophisticated than others.[72] The assumptions that go into exposure models mostly reflect the modelers' best judgment as to how pollutants and people interact in the real world, but they can also reflect policy judgments about whether to err on the side of over- or under-prediction of exposure.[73]

The same is true for the models that the staff employs to assess the risks to various human subpopulations. Since scientists disagree about the assumptions that should drive predictive models, different models can yield vastly different predictions.[71] In choosing from the available models, the agency staff again is making a policy judgment about the extent to which the agency should err on the side of safety by choosing a model that may overpredict risk or err on the side of the status quo by choosing a model that may underpredict risk.[74]

At this applied stage, the scientific judgments in the Risk/Exposure Assessment thus become even more pronounced than was the case in the ISA.[17,73] The dangers of bias in the collective scientific reports are again checked in part by another round of rigorous peer review by a science advisory board as well as by public comment.[21] This vetting of this second report again helps (although it does not ensure), that contestable judgments made by the scientific authors are drawn into the light and qualified in a scientifically rigorous way.

The third and final "firewalled" science-intensive report prepared by the technical staff at EPA—called the "Policy Assessment"—seeks to translate the prior two scientific reports for policy makers in a policy-analysis report.[21] The translation effort is again subject to scientific peer review as well as public comment and as a result, most of the value judgments are presumably drawn out and qualified and/or adjusted over time. Attribution of authors in all three reports also helps to encourage and reward high quality analyses.[21]

Once these three reports have been finalized, the scientific record is considered complete. Policy makers can then use or ignore this information as they choose. It is not uncommon for the EPA Administrator to occasionally select ambient standards that are not in accord with the range of plausible options detailed in these scientific assessments.[75] In some cases, the officials' politically preferred standards may even be so poorly explained that the standards are later reversed by a court in a legal challenge brought by stakeholders.[75] In this way, scientific findings lie at the base of the policy decision and cannot be manipulated or ignored.

Certainly, this kind of elaborate multi-step process cannot work for all agency science-policy decisions. And it will not work in emergency settings, like COVID, where swift decisions are imperative. But the underlying goal of isolating rigorous consensual scientific advice and placing it in the public record to act as a constraint on subsequent political decision-makers is nonetheless a valuable structural addendum to protect against the worst forms of politicization. At the same time, ensuring that the internal scientific advice is rigorously scrutinized by peers and the general public helps protect against scientific judgments that otherwise might be black-boxed as inscrutable scientific advice.

The future viability of implementing this structured approach across the government is, not surprisingly, impeded by—what else but politics! As a practical matter, political appointees appear loathe to give up the luxury of occasionally controlling what the science says. They may thus avoid adopting these kinds of measures voluntarily. Even Congress may find that its own political power is more limited when the science cannot be politicized. Thus, the ultimate obstacle to this type of institutional design may regrettably fall back to the fact that the fox (politics) is ultimately in charge of the hen house (science).

A second challenge with this reformed approach lies with the difficulty of developing a realistic way to insulate expert staff from at least some political control over hiring, management, peer reviewers, and budgets. The revised NAAQS approach just discussed does offer some valuable structure and barriers to the politicization of science during the decision process itself. But under that approach, the scientific staff still remains vulnerable to budget cuts, hiring freezes, and management changes designed to limit staff independence. And political officials still select the members of the Science Advisory Committee that peer reviews the agency's work.[14] Indeed, the EPA Administrator under President Trump further adjusted the line-of-command so that EPA's NAAQS scientific office was required to report to a political appointee, rather than the Chief Scientist, who was the direct supervisor of that staff over the preceding decades.[25]

5. Conclusion

The politicization of science is a recurring phenomenon in US federal policymaking that is explained in part by the unstructured, collaborative nature of decision-making in most science-intensive US regulatory programs. In this chapter we have endeavored to spotlight some of the most significant worries arising from this longstanding approach to U.S. institutional design for expert agencies and offer recommendations for stronger institutional structures that should guard against some of the risks of politicized science. We are cautiously optimistic that precisely because the agency work is so important and the problems are so pervasive, there will be opportunities for reform ahead.

References

1. Michaels D. *Doubt Is Their Product: How Industry's Assault on Science Threatens Your Health*. Oxford, England: Oxford Press; 2008.
2. Michaels D. *The Triumph of Doubt Oxford*. England: Oxford Press; 2020.
3. Oreskes N, Conway E. *Merchants of Doubt: How a Handful of Scientists Obscured the Truth on Issues From Tobacco Smoke to Global Warming*. New York, NY: Bloomsbury Publishing; 2009.
4. McGarity T, Wagner W. *Bending Science: How Special Interests Corrupt Science Health Research*. Cambridge, MA: Harvard University Press; 2007.
5. Legg T, Hatchard J, Gilmore A. The science for profit model—How and why-corporations influence science and the use of science in policy and practice. *PLoS ONE*. 2021;16(6): e0253272. https://doi.org/10.1371/journal.pone.0253272.
6. ICJME. *Recommendations for the Conduct, Reporting, Editing, and Publication of Scholarly Work in Medical Journals*; 2019. http://www.icmje.org/icmje-recommendations.pdf.
7. Holman B, Elliott K. The promise and perils of industry-funded science. *Philos Compass*. 2018; e12544.
8. Munafo M, et al. A manifesto for reproducible science. *Nat Hum Behav*. 2017;1:1.
9. Coglianese C. The semi—Autonomous administrative state. In: *The Regulatory Review*; Oct 7, 2019. https://www.theregreview.org/2019/10/07/coglianese-semi-autonomous-administrative-state/.
10. OECD. *The Governance of Regulators' Practices: Accountability, Transparency and Co-ordination*. Paris, France: OECD Publishing; 2016.
11. OECD. *Being an Independent Regulator*. Paris, France: OECD Publishing; 2016.
12. Bharara P et al., Brennan Center, National Task Force on Rule of Law & Democracy Volume II, Proposals for Reform. https://www.brennancenter.org/our-work/policy-solutions/proposals-reform-volume-ii-national-task-force-rule-law-democracy. Published 2019.
13. Bipartisan Policy Center. Improving the Use of Science in Regulatory Policy. https://bipartisanpolicy.org/download/?file=/wp-content/uploads/2019/03/BPC-Science-Report-fnl.pdf. Published 2009.
14. Goldstein B. What the trump administration taught us about the vulnerabilities of EPA's science-based regulatory processes: Changing the consensus processes of science into the confrontational processes of law. *Health Matrix*. 2021;31:299.

15. Wagner W. A place for agency expertise: Reconciling agency expertise with presidential power. *Columbia Law Rev.* 2015;115:2041.
16. Committee on the Institutional Means for Assessment of Risks to Public Health, National Research Council of the National Academies of sciences. *Risk Assessment in the Federal Government: Managing the Process.* Washington, DC: National Academies Press; 1983.
17. McGarity T. Substantive and procedural discretion in administrative resolution of science policy questions: Regulating carcinogens in EPA and OSHA. *Georgetown Law J.* 1979;67:729.
18. Doremus H. Scientific and political integrity in environmental policy. *Tex Law Rev.* 2008;86:1601.
19. Reilly-Diakun J. Addressing blurred lines: Institutional design solutions to transgressions across the science-policy boundary. *Tex Environ Law J.* 2019;49:199.
20. Kitrosser H. Scientific integrity: The perils and promise of white house administration. *Fordham Law Rev.* 2011;79:2395.
21. Wagner W. Report for ACUS: Science in Regulation: A Study of Agency Decisionmaking Approaches. https://www.acus.gov/sites/default/files/documents/Science%20in%20Regulation_Final%20Report_2_18_13_0.pdf. Published 2013.
22. 5 U.S.C. § 552(b)(5) n.d.
23. Narayan S. Proper assertion of the deliberative process privilege: The agency head requirement. *Fordham Law Rev.* 2008;77:1183.
24. Memoranda from President Barack Obama to the Heads of Exec. Dep'ts & Agencies, Mar. 9, 2009, and from John P. Holdren to the Heads of Exec. Dep'ts & Agencies, Dec.17, 2010.
25. McGarity T, Wagner W. Deregulation using stealth science strategies. *Duke Law J.* 2019;68:1719.
26. Steinzor R. *Mother Earth and Uncle Sam: How Pollution and Hollow Government Hurt Our Kids.* New York, NY: Cambridge Press; 2008.
27. Metzger G. Foreward: 1930s redux: The administrative state under siege. *Harv Law Rev.* 2017;131(1):90.
28. Steinbrook R. Science, politics, and Federal Advisory Committees. *New Engl J Med.* 2004;350:1454.
29. U.S.C. § 5(b)(2) n.d.
30. U.S.C. § 5(b)(2) n.d.
31. Perls H. Deconstructing environmental deregulation under the trump administration. *Vermont Law Rev.* 2021;45:591.
32. Tracy A. How trump gutted Obama's pandemic-preparedness systems. *Vanity Fair.* May 1, 2020.
33. Qui L, Marsh B, Huang J. The president vs. the experts: How Trump downplayed the coronavirus. *New York Times.* March 20, 2020; A13.
34. Lewis M. *The Premonition: A Pandemic Story.* New York, NY: W.W. Norton Publisher; 2021.
35. Slavitt A. *Preventable: The Inside Story of How Leadership Failures, Politics, and Selfishness Doomed the Coronavirus Response.* New York, NY: St. Martins Press; 2021.
36. McGinley L. Dozens of coronavirus antibody tests on the market were never vetted by the FDA, leading to accuracy concerns. *Wash Post.* April 19, 2020.
37. Eder S, Twohey M, Mandavilli A. Flaws hinder antibody testing. *New York Times.* April 20, 2020; A1.
38. Burton T. FDA sets antibody testing standards. *Wall Street J.* May 5, 2020; A4.
39. Mullis A, Gardenswartz A, Eiselein E. Missteps to avoid in marketing COVID-19 antibody tests. *Law.* May 11, 2020;360.
40. McGinley L. FDA steps up scrutiny of coronavirus antibody tests to ensure accuracy. *Wash Post.* May 5, 2020.

41. Roland D. Roche antibody test wins FDA approval for emergency use. *Wall Street J.* May 4, 2020;A8.
42. Weixel N. FDA grants authorization to new abbot COVID-19 antibody test. *The Hill.* May 11, 2020. https://thehill.com/policy/healthcare/497114-fda-grants-authorization-to-new-abbott-covid-19-antibody-test.
43. Abutaleb Y, McGinley L, Dawsey J. Oracle to partner with trump administration to collect data on unproven drugs to treat covid-19. *Wash Post.* March 24, 2020.
44. Katz, Marshall & Banks, n.d. Addendum to the Complaint of Prohibited Personnel Practice and Other Prohibited Activity by the Department of Health and Human Services Submitted by Dr. Rick Bright. https://kmblegal.com/sites/default/files/NEW%20R.%20Bright%20OSC%20Complaint_Redacted.pdf.
45. Weiland N, Haberman M. Oracle is enlisted by the white house to aid in the study of unproven drugs. *New York Times.* May 25, 2020; A9.
46. Thomas K, Grady D. Embrace of a drug goes against science. *New York Times.* March 21, 2020; A13.
47. Rowland C, Johnson C, McGinley L. Trump calls anti-malarial drug a "game changer" for coronavirus, but the FDA says it needs study. *Wash Post.* March 19, 2020.
48. Busari S, Adebayo B. Nigeria records chloroquine poisoning after trump endorses it for coronavirus treatment. *CNN World.* March 23, 2020.
49. Parkinson J, Gauthier-Villars D. Trump sparks run on two drugs. *Wall Street J.* March 24, 2020; A9.
50. Grady D, Kannapell A. Trump urges use of drug yet unproven. *New York Times.* April 5, 2020;A10.
51. Dawsey J. "What do you have to lose?": Inside Trump's embrace of a risky drug against coronavirus. *Wash Post.* April 7, 2020.
52. Page J. Chinese doctors question sought-after drug. *Wall Street J.* April 10, 2020; A16.
53. Osborne H. French hospital stops hydroxychloroquine treatment for COVID-19 patient over major cardiac risk. *Newsweek.* April 8, 2020.
54. Food and Drug Administration, Coronavirus (COVID-19) Update: FDA Reiterates Importance of Close Patient Supervision for "Off-Label" Use of Antimalarial Drugs to Mitigate Known Risks, Including Heart Rhythm Problems. April 24, 2020. https://www.fda.gov/news-events/press-announcements/coronavirus-covid-19-update-fda-reiterates-importance-close-patient-supervision-label-use.
55. Editorial. The results are in Trump's miracle drug is useless. *Wash Post.* May 18, 2020.
56. Blake A. FDA's hydroxychloroquine reversal raises even bigger questions about Trump's role in pushing for the drug. *Wash Post.* June 15, 2020.
57. Kopp E. States weigh what to do with millions of malaria pills. *Roll Call.* 2020.
58. Timm J. Trump Says He's no Longer Taking Hydroxychloroquine. NBC News.com. May 25, 2020. https://www.nbcnews.com/politics/donald-trump/trump-says-he-s-no-longer-taking-hydroxychloroquine-n1214301.
59. Baker P. Trump and Kushner boast of success in recasting administration's response. *New York Times.* April 30, 2020; A8.
60. Chalfant M. White house officials downplay chance of COVID-19 "second spike.". *The Hill.* June 12, 2020.
61. Sanmez F. Trump's Tulsa campaign rally sign-up page includes coronavirus liability disclaimer. *Wash Post.* June 12, 2020.
62. Wright L. The plague year. *New Yorker.* 2021;20.
63. Centers for Disease Control, n.d. Covid Data Tracker Weekly Review, https://www.cdc.gov/coronavirus/2019-ncov/covid-data/covidview/index.html.
64. Evans M, et al. Covid-19 Hospitalizations Approach a Peak. *Wall Street J.* August 31, 2021;A7.
65. Physicians for Social Responsibility v. Wheeler, 956 F.3d 634 (D.C. Cir. 2020) n.d.

66. Wagner W, et al. Public participation in agency rulemakings: An empirical study. *Adm Law Rev*. 2021;73(1):609.
67. McGarity T. *Freedom to Harm*. New Haven, CT: Yale University Press; 2013.
68. 42 U.S.C. § 7409(b)(1) n.d.
69. 42 U.S.C. § 7408(a)(2) n.d.
70. EPA. *Preamble to the Integrated Science Assessments*; 2015. https://cfpub.epa.gov/ncea/isa/recordisplay.cfm?deid=310244.
71. McGarity T. A cost-benefit state. *Adm Law Rev*. 1998;50:7.
72. Watnick V. Risk assessment: Obfuscation of policy decisions in pesticide regulation and the EPA's dismantling of the food quality protection act's safeguards for children. *Ariz St L J*. 1999;31:1315.
73. McGarity T. Science and policy in setting national ambient air quality standards: Resolving the ozone enigma. *Tex Law Rev*. 2015;93:1783.
74. Latin H. Good science, bad regulation, and toxic risk assessment. *Yale J Regul*. 1988;5:89.
75. Shapiro S, Fisher L, Wagner W. The enlightenment of administrative law: Looking inside the agency for legitimacy. *Wake Forest Law Rev*. 2013;47:463.

CHAPTER TWO

Beyond the sheltering academic silo: Norms for scientists' participation in policy

Karen L. Akerlof*
Department of Environmental Science & Policy, George Mason University, Fairfax, VA, United States
*Corresponding author: e-mail address: kakerlof@gmu.edu

Contents

1. Introduction 29
2. Protective silos: Science and objectivity 31
3. The evolving relationship between science and policy 32
4. Norms for engaging in hyperpolitical settings: Legislatures 35
5. Interviews with congressional staff regarding appropriate scientist roles in policy 36
6. Re-imagining scientists' engagement with policy 40
7. Conclusion 41
References 41

Abstract

Over the past 75 years, the relationship between science and policy has been transformed as U.S. government priorities shifted from basic research post-World War II toward scientific knowledge that contributes to societal decision-making and spurs innovation. These macrolevel forces have restructured the landscape in which scientists interact with policymakers, creating new norms for their involvement in policy. This chapter focuses specifically on legislatures as one of the most consequential fora for policy decisions, but also hyperpolitical because of its representative function in democracies. In interviews, legislative staff in the U.S. Congress, who serve as gatekeepers for the flow of policy-relevant information, embrace a wide range of norms for scientists' involvement, but they also cite concerns about the effects of redefining these boundaries on scientists' authority. Researchers and their institutions should be aware of these broad trends in conceptualizing and strategizing their approaches to societal impact.

1. Introduction

In spring 2017, approximately 100,000 scientists joined the March for Science on Washington, DC in an "unprecedented" protest of the treatment of science under the Trump administration.[1,2] While academics have since

Progress in Molecular Biology and Translational Science, Volume 188
ISSN 1877-1173
https://doi.org/10.1016/bs.pmbts.2021.11.007

criticized the march for potentially contributing to the political polarization of science,[3,4] this event served as a visible representation of a much broader set of trends that have been occurring for decades, reordering the relationship between science, policy, and society. In this chapter, I describe how the role of science vis a vis society has changed over decades in dramatic ways that inherently make it more political, but that scientists may not necessarily realize or acknowledge.[5] The politicization of science and the scientization of policy serve as opposite sides of the same coin.[6] This chapter focuses specifically on legislatures as sites for science advice to decisionmakers. These chambers represent hyperpolitical fora,[7] further elevating inherent tensions between expert knowledge and democratic representation.[8] In interviews, legislative staff in U.S. Congress—who serve as gatekeepers for the flow of policy-relevant information—embrace a wide range of norms for scientists' involvement, but they also cite concerns about the effects of redefining these boundaries on scientists' authority.[9] While the evidence presented in the chapter is not specific to any particular disciplinary field, molecular biology and other related fields are no stranger to the phenomena described here.[10,11] How well scientists and research institutions adapt to these new conditions will depend on their understanding of these macrolevel shifts (Fig. 1).

Fig. 1 In spring 2017, Trump administration policies spurred protests over the role of science in policy. *Photo credit: author*

2. Protective silos: Science and objectivity

Among the professed ideals of modern science is objectivity, in the sense of revealing true knowledge about the state of nature.[12] But only relatively recently in the history of science—the mid-nineteenth century—did objectivity emerge as a subject of concern.[13] Whereas earlier scientists concerned themselves with the epistemological challenges posed by the variability of nature, by circa 1860 scientists perceived another challenge to their ability to describe and explain nature: that the subjective self was prone "to see what it wanted to see" (p. 34). One way to address the perils of individual scientists' subjectivity is through what Merton termed the collective scientific norm of "disinterestedness."[14] The institutionalization of this norm offered the potential protection of the scientific enterprise from the temptation to make untestable claims, and the public against the abuse of expert authority. Conversely, the practices of disciplinary communities served to shield their fields from social and political influence.[12]

The idea that science itself is value-free, not influenced by nor relevant to questions of ethics and morals, dates at least to the turn of the twentieth century writing of sociologist Max Weber[15,16] and still has adherents today.[17] Yet, as scholars starting in the mid-nineteenth century brought to light, values do not only come into play the moment that scientists exit the laboratory door. In the 1970s, researchers focused on the social contexts of scientific disciplines. Kuhn suggested that the beliefs and practices of academic communities follow somewhat arbitrary paths set by history and accident that provide the structure of their rules, practices, and associated knowledge—e.g., paradigms—that allow their adherents to efficiently frame and answer scientific questions.[18] The idea that the production of scientific knowledge is a social process was pushed even further by sociologists like Latour who, in an ethnographic account of the activities of bench researchers at the Salk Institute, demonstrated the fundamental role of value judgments in the production of knowledge, a factor that becomes obscured when scientific findings are reported. He thus labeled the process of science the "social construction of scientific facts."[19(p32)] Starting with the selection of the research problem and extending to data collection, acceptance of a theory or hypothesis, and finally the application of the findings,[20] each stage of the process requires scientists to make judgments either based on their own experiences or disciplinary norms.[18] Thus knowledge is shaped by the structure of scientific communities, organizations that themselves can be considered political because they help constitute the fabric of societal culture and power.[21(p21)]

Indeed, Douglas and other philosophers of science have posited that the practice of science *should not* be value-free, in that values play a role in science across all its stages and that they do not necessarily diminish its objectivity.[22] Douglas describes values as either epistemic—the judgments that are made within disciplines that guide the process of knowledge discovery in ways that increase its reliability and accuracy—or ethical, social, and cognitive, requiring judgments related to the implications of research outside a discipline. Ethical values help researchers weigh the potential harm from scientific errors to the public; social values may direct them to study one area over another because of its potential to contribute to the public good; and cognitive values including simplicity and explanatory power contribute to the ability of other researchers to build on the work.

Non-epistemic values like these should play a role in scientific research, according to Douglas,[22] because often decisions are made in situations of uncertainty, where scientific evidence alone cannot provide an answer. In these cases, non-epistemic values should play an indirect role along with the scientific evidence. Values are only incompatible with science when they serve as the *sole* motivation for decisions, writes Douglas.

3. The evolving relationship between science and policy

The philosophical recognition that values play a role in scientists' decision-making—and that scientific communities are political entities—would remain a fairly esoteric epistemological issue if it weren't for the evolution of the relationship between science and policy in the United States since World War II. In 1944, President Franklin D. Roosevelt penned following words[23]: "New frontiers of the mind are before us, and if they are pioneered with the same vision, boldness, and drive with which we have waged this war we can create a fuller and more fruitful employment and a fuller and more fruitful life." In response, Vannevar Bush's proposal to the president,[23] and a similar report a few years later,[24] argued for the large-scale government funding of research and the establishment of what would eventually become the National Science Foundation. In *Science: The Endless Frontier*, research fell into three categories—pure, background, and applied—but only the first was prioritized.[23] While pure or basic[25] research is "without specific practical ends" (p. 75), applied research is conducted for the purpose of achieving a specific outcome of anticipated value. Bush also described a third type of research as the process of collecting data and establishing standards used within both pure and applied studies, or "background" science. In recommending governmental policies for funding science during peacetime, Bush contended (p. 77): "... applied research invariably drives out pure. The moral is clear: It is pure research which deserves and requires special protection and specially assured support."

Gibbons and colleagues[26] describe the post-war focus on *policy for science* as one of three historical stages that contribute to the evolving conceptualization of science and technology (S&T) policy. (These periods roughly align with those described by other scholars.[27(p141)]) As agencies bid for higher levels of research funding in U.S. congressional appropriations bills every year, *policy for science* discourses continue to this day. In FY 2021, the annual budget for the National Science Foundation was $8.5 billion and the National Institutes of Health, $42.9 billion.[28] What has changed, however, is the expectation for what the government receives in exchange. Increasingly federal funding agencies stipulate that researchers will conduct science that has societal impact. Between 2000 and 2017, the share of U.S. federally funded research dollars for basic research declined from 58% to 42%, while applied research rose from 27% to 35%.[29]

Famously distinguished from the post-war *policy for science* by Brooks,[30] *science for policy* became an increasing focus in the 1970s during a period of economic crisis. A 1971 OECD report captured the prevailing winds: "A crisis of growth and a crisis of the linkage of science policy to social aspirations seem to be two closely related aspects of the same problem: the aim of science policy cannot be solely the expansion of S&T but must include the management and direction of technological progress for the service of other ends."[31(p460)] During the 1990s, in a third shift, the end goal in supporting S&T was further redefined by OECD as "innovation," which could be directly connected to markets through its implementation.[32]

As this abbreviated history demonstrates, within a period of approximately 50 years, governments' thinking about the relationship between science and government transformed, not just in the United States but internationally. In the aftermath, Stokes[33] went so far as to argue that pure science could no longer be presumed a public good:

> *.. vigorous public support for basic research is unlikely to be restored by asserting the case for pure science, in accord with the postwar paradigm. Indeed, this assertion increasingly casts the scientific community in the role of an interest group seeking support for an activity that reflects its own essential needs rather than in the role of informed spokesman for a general interest. However appealing the scientific community may find the ideal of pure inquiry and the autonomy it brings, these do not offer a clear and powerful case for public support of basic science. (p. 96)*

The ways that research is conducted have shifted, too, in accommodation. New models have arisen to conceptualize and describe the implementation of science that no longer sits within academic silos. Moving beyond the dichotomy of "pure" and "applied," Stokes proposed "use-inspired" basic research.[33] His classification of research across two dimensions—consideration of use and a quest for understanding—distinguished between applied research that did not seek to contribute to an understanding of nature and research that had both practical and fundamental goals. Taking an alternate approach that focuses on the social processes inherent to these new ways of conducting science, Gibbons and colleagues described "Mode 2" production of knowledge as the alternative to the "Mode 1" model of pure science.[26] Mode 2 research represents a form of applied science that is transdisciplinary, produced for a specific application, conducted by diverse groups of people, socially accountable, and held to quality standards defined by its audiences, not those of scientists alone. Similar descriptions of this complex relationship between knowledge production and diverse publics appear within a wide array of academic literatures spanning domains of policy and practice, but under different names.[34] Examples include participatory research,[35] action research,[36] civic science,[37] community science,[38] post-normal science,[39] and co-production.[21,40]

A third modality—"Mode 3"—further widens knowledge production to economic and social systems spanning broad groups, sites, and modes across multiple levels.[41,42] Within this context, the ecosystem for innovation spans universities, industry, government, and the media- and culture-based public.[43] Within the White House Office of Science and Technology Policy (OSTP), the appointment of the first Deputy Director for Science and Society, Dr. Alondra Nelson, in 2021 serves as an indicator for the extent to which a broader vision for inclusion holds sway in the U.S. government. "The goal is to build a science policy that intentionally and explicitly includes the perspectives of the American public, including seeing science and technology through the eyes of folks who are marginalized or vulnerable," according to Nelson,[44] in acknowledgement that while science has served as a source of "great progress and promise" to some communities, it has come at a cost to others.

Indeed, the rapidity and scope of scientific and technological change[45] puts pressure on societies worldwide to reconceptualize its governance. The challenge for researchers in this "participatory turn"—as termed by Jasanoff—is that it is the scientific enterprise's very traditions of gatekeeping that have enabled it to present as the provider of "truth."[46,47] Scientific communities establish boundaries, delineating who has professional credibility within their disciplines and what is—or is not—authoritative knowledge. And they engage in practices, such as publication and peer review, to control the process of knowledge production and ensure it aligns with those standards. As these self-imposed boundaries erode through the scientization of policy—and politicization of science—they expose scientific knowledge to deconstruction within political processes and scientists themselves to the need to defend their expert status within decision-making and democratic governance.

4. Norms for engaging in hyperpolitical settings: Legislatures

The pervasiveness of science and technology within society implicates its use for policy not only at all levels of governance—local, state, federal, international—but within executive agencies, legislatures, and even the judicial branch. Legislatures—by their very nature—serve as one of the most hyperpolitical sites for S&T advice because of their representative function within democracies. As a result, they require unique models for its delivery.[48,49] In the United States, the creation of the Office of Technology Assessment (OTA) in the 1970s to inform Congress as part of government reform efforts served to inspire the establishment of similar bodies in other countries.[50,51] But the creation of these organizations for the purpose of assisting legislatures in addressing policy issues of societal importance also has continued to raise questions about the fundamental nature of scientific knowledge and its use in governance.[52,53] Even organizations like OTA—and now GAO's Science, Technology Assessment, and Analytics Team—that perform the role of "boundary organizations" in legislatures, mediating between various scientific and policy communities and providing different types of accountability to each,[54] can be subject to perceptions of politicization that threaten their long-term survival.[49] Scientists themselves are most likely to participate in policy processes through direct involvement, either by contacting their elected representatives or agency staff, as opposed to working through these organizations that bridge the science and policy communities. A survey of American Association for the Advancement of Science members found that more than half (55%) had communicated with a policymaker in the previous 2 years, but the majority had not worked with a boundary organization (68%).[55]

Given how the changes over the last several decades in norms for scientific involvement in policy described in the preceding sections, my colleagues and I sought to understand how legislative science advice experts worldwide—science advice researchers, providers of scientific information to government, and users of scientific information within government—viewed the appropriate role of scientists in policy.[56] (The research methodology is described in full in Akerlof et al.[48]) In 2018, we asked them, "Which of the five roles below do you think are most helpful in informing legislative policy processes?" The experts could select one or more of the options adapted from Steel et al.[57]:

1) Scientists should only report scientific results and leave others to make policy decisions.
2) Scientists should report and then interpret the results for others who are involved in policy decisions.
3) Scientists should work closely with policymakers and others to integrate scientific results in policy decisions.
4) Scientists should actively advocate for specific policies they prefer.
5) Scientists should be responsible for making decisions about policy.

About half of the respondents ($n = 183$) identified as from developing nations and half from those considered developed.[58] We found that the norms of scientists to take more active roles in informing policy had been largely adopted by international experts, even within hyperpolitical contexts such as legislatures. The most commonly selected "appropriate" roles were (1) that scientists should work closely with policymakers and others to integrate scientific results in policy decisions (79.2%) and (2) that they should report and translate results (53.0%). The traditional role of "pure" scientists—just reporting results—was not as likely to be perceived as an appropriate role (12.0%), but neither were advocacy (18.6%) or fully participatory roles in policymaking (11.5%).

5. Interviews with congressional staff regarding appropriate scientist roles in policy

Arguably, the U.S. Congress is even more hyperpolitical than other similar institutions due to the recent era of razor-thin party majorities.[7] In 2017, we conducted structured interviews with 16 staff members who were employed in the personal offices of members of the U.S. Congress.[9] The House and Senate each have an energy, natural resources/environment, and science committee. The staff members who support the work of the members of those committees served as the population for the study because of the relevance of science to their portfolios and their role as informational gatekeepers and decision-makers within the institution.

The study sample recruitment was stratified to achieve roughly equal proportions from both chambers and political parties: House Democrat, House Republican, Senate Democrat, and Senate Republican. The recruitment list within each of these groups was randomized; potential respondents were contacted systematically by phone and email until the quota of about 4 per group was achieved. (One of the interviewees did not answer the question about appropriate.

roles for scientists in policy due to time constraints.) The relatively small sample number was chosen for two reasons: (1) to conduct longer, more in-depth interviews to explore the full extent of the study's research questions; and (2) to restrict sampling of the limited study population for a subsequent research project.

Using the same categorization of "appropriate roles" as in the previous study, we asked: "Which of the five roles do you think is most helpful in informing policy processes?" and "Why?" All but one staff member supported more active roles in policy for scientists than just reporting their research findings. Again, as in the previous study, respondents were most supportive of scientists working closely with policymakers to integrate scientific results in policy decisions (10), followed by reporting and then interpreting the results for policymakers (9), and advocating for specific policies (7). We asked the staff members to pick one role as more helpful than others, but most (9 of 15) chose multiple roles as equally helpful (Fig. 2), with four saying that all of the roles were appropriate. While staff members were generally supportive of scientists playing a wide array of roles in policy decisions, they also drew distinct boundaries between the authority of the science and policy communities. In describing "why" they chose each set of categories as appropriate roles for scientists, the staff members highlighted three areas of tension. They demarcated science's limitations to influence policy within the political sphere, acknowledged their presumption of bias and advocacy within the legislative context, and highlighted the risk of politicization to scientists by adopting advocacy roles. Each of these three topics—and associated example quotes—are presented below.

Science alone can't drive policy. Staff members called attention to the different types of value judgments and knowledge that are needed in policy processes, and that scientific information was not the most important among considerations. While these comments fall in line with the conclusions of

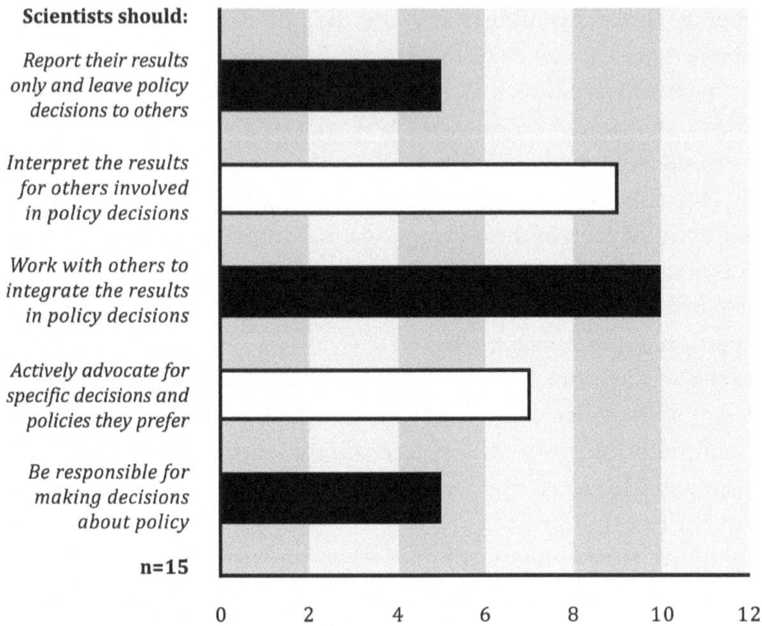

Fig. 2 In interviews, congressional staff said scientists should integrate research findings with policy.[9]

decision analysis scholars,[59] they also demonstrate the limited potential effect of calls for policymakers to "listen to science."[60] Staff statements suggest some of the difficulty in reaching mutual understanding for how to achieve complementarity between the two communities.

- As un-versed in science as most policymakers are, most scientists are that ill-versed in policy. ...If you don't understand policy, then you don't understand policy. That's our factory floor over there, and we do what we do, and there's a lot of inputs. Science is one of those. But like I said, I have a really hard time finding a policy that was required because of a specific scientific result.
- So science is explaining how the system works, that doesn't say whether we should turn left or turn right. It just says if you turn the steering wheel one way, we're going to go that way. Almost everything we do up here is a value judgment about like, all right, we should turn right or we should turn left, which isn't really ... as long as you tell me which way I need to turn the steering wheel to go left or right, you leave it up to the elected officials to say we should go left or right. So I think it's that interpreting, providing sort of here's how things work, now you decide which way we need to go.

Everyone lobbying the Hill has a policy bias; scientists should also be advocates. Some communication scholars define the lack of bias as a prerequisite for credibility in persuasion.[61] Yet, congressional staff members can name few sources of scientific information that they consider unbiased.[9] Building credibility in this context may thus require different considerations, for example, in the transparency with which scientists convey their own values in juxtaposition to others within the policy issue space. Indeed, similar recommendations have been made by the European Commission.[62]

- ... I certainly would prefer for them to be advocates for positions. I think that's more helpful because ... I mean regardless of who they are and what the basis is for their decision ... I know that everybody has a position on a policy, and I would rather that somebody state their position or their organization's position, and tell me whose position it is and why, than for them to attempt to maintain some degree of objectivity... when I said earlier that I spent a lot of time trying to assess the credibility of information, part of that is looking at who's providing the information and looking at what perspective they're coming from. And so I'm gonna be doing that anyway and so I think it gets to be more complicated if somebody is trying to mask their perspective and providing what they perceive to be objectivity.
- The good [lobbyists] are the one who say, 'I'm here today on behalf of this client, let me also bring you some information, that isn't entirely consistent with their position.' That is how you build huge amounts of credibility. It is perfectly acceptable to me, and I would welcome if I knew the scientist well enough to have him advocate for a policy position. Because to say, here is the evidence, let me help you do the work of sorting through these issues, because we are not staffed as we used to be to develop the in-house expertise. I'm stretched way too thin to do that, so it would be helpful.

Crossing the advocacy line diminishes the authority of science. Staff members also voiced caution that in crossing into the policy sphere, scientists would potentially endanger the credibility of their previous research if it became associated with a political agenda. These arguments reflect longstanding concerns by scholars that the boundary work between the communities serve a purpose in protecting their respective authority.[54] The value in presenting the "ready-made science" of established facts to policymakers[19] is that it reduces their ability to deconstruct the knowledge in ways that may be amenable for their political purposes.

- [Scientists should say ...] here's what the facts say. Leave it to somebody else to determine what does that mean and how should we respond to that. It kind of keeps it above the political fray... and doesn't risk lumping the outcomes of the research in with the more partisan happenings here and how that should be addressed. If I'm just a scientist and I've done a ton of climate change research and suddenly I'm here advocating for this policy, I have a lot of credibility because of that, but there's a risk that maybe my thoughts on other issues or even my policy work here, and the tactics I choose to address that, might drown out some of the efficacy of the work I've done.
- A side effect of having scientists being the ones lobbying is that science then loses its allure as unpolitical. Science is already politicized enough. Policymakers need to have better access to scientific research in formats that are access(ible), but I don't necessarily think that every scientist needs to be a lobbyist or an advocate, too.

6. Re-imagining scientists' engagement with policy

Some have argued that all highly policy-relevant science is politicized (and biased),[63] and indeed, evidence suggests that as policy disputes become more adversarial, the authority of science plays an increasingly important role.[64] This brief history of the evolution of S&T policy priorities and description of the views of legislative experts and staff intimates that scientists will increasingly find themselves outside of academic silos, engaging in policy issues that are scientized and in which science has been politicized. Yet, as other scholars have suggested, there are alternative ways that researchers and their institutions can engage in these landscapes that may offer some protection from political vicissitudes. For example, boundary organizations can help increase access to information to broader audiences of decision-makers while allowing researchers to focus on the production of knowledge.[65] Universities are increasingly developing and promoting these capabilities in-house to connect researchers to wider societal ecosystems of knowledge, innovation, and governance.

But for researchers and their institutions to take advantage of these types of mechanisms, they have to first comprehend the extent to which the relationship between science and policy has changed within the last 75 years, along with the evolution of new norms for scientists' engagement. Science and technology have never been so important to society globally, a state of affairs that is unlikely to change in the near term. Given its significance to the lives of everyday citizens and their collective governance, the trajectory for science and technology is more open, participatory … and yes, political. As Eric Lander, the President's Science Advisor and Director of the White House Office of Science and Technology Policy, noted, "We need everybody to be able to participate in and contribute to S&T because different experiences and perspectives are the bedrock of new scientific and technological insights, because having everybody on the team is essential to America's global competitiveness in the 21st century, and, because it's the right thing to do."[66]

7. Conclusion

This chapter describes the forces behind boundary shifts in the science policy landscape and how legislative experts and staff in the U.S. Congress conceive the resulting norms. In drawing attention to long-term macrolevel forces that shape how science and technology relate to society, it points to the need for researchers and their institutions to strategically develop mechanisms that promote their ability to produce and convey societally relevant knowledge. As higher education re-imagines itself in the coming decades, its relationship to innovation and governance will require that it also rethink how it connects to societal ecosystems for knowledge creation, dissemination, and use.

References

1. Appenzeller T. An unprecedented march for science. *Science*. 2017;356(6336): 356–357. https://doi.org/10.1126/science.356.6336.356.
2. Fleur NS. *Scientists, Feeling under Siege, March against Trump Policies*. The New York Times; 2017. https://www.nytimes.com/2017/04/22/science/march-for-science.html. Published April 22, Accessed November 2, 2021.
3. Brulle RJ. Critical reflections on the march for science. *Sociological Forum*. 2018;33 (1):255–258. https://doi.org/10.1111/socf.12398.
4. Motta M. The polarizing effect of the march for science on attitudes toward scientists. *APSC*. 2018;51(4):782–788. https://doi.org/10.1017/S1049096518000938.
5. Turnhout E. The politics of environmental knowledge. *Conserv Soc*. 2018;16 (3):363–371.
6. Weingart P. The scientific power elite—A chimera; the de-institutionalization and politicization of science. In: *Scientific Establishments and Hierarchies*. Springer; 1982:71–87.
7. Lee FE. *Insecure Majorities: Congress and the Perpetual Campaign*. The University of Chicago Press; 2016. https://press.uchicago.edu/dam/ucp/books/pdf/course_intro/978-0-226-40904-7_course_intro.pdf.
8. Guston D. The essential tension in science and democracy. *Soc Epistemol*. 1993;7 (1):3–23. https://doi.org/10.1080/02691729308578676.
9. Akerlof K. *Congress's Use of Science: Interviews with Energy, Natural Resources, and Science Personal Office Staff Members*. George Mason University; 2019.

10. Wright S. Molecular biology or molecular politics? The production of scientific consensus on the hazards of recombinant DNA technology. *Soc Stud Sci*. 1986;16(4):593–620. https://doi.org/10.1177/030631286016004003.
11. Wright S. Molecular politics in a global economy. *Politics Life Sci*. 1996;15(2):249–263.
12. Porter TM. *Trust in Numbers*. Reprint edition. Princeton University Press; 1996.
13. Daston L, Galison P. *Objectivity*. Zone Books; 2010.
14. Merton RK. *The Sociology of Science: Theoretical and Empirical Investigations*. University of Chicago Press; 1973.
15. Sharlin AN. Max Weber and the origins of the idea of value-free social science. *European J Sociol*. 1974;15(2):337–353.
16. Weber M. *The Methodology of the Social Sciences*. Free Press; 1949.
17. Betz G. In defence of the value free ideal. *Euro Jnl Phil Sci*. 2013;3(2):207–220. https://doi.org/10.1007/s13194-012-0062-x.
18. Kuhn TS. *The Structure of Scientific Revolutions*. 2nd edition. University of Chicago Press; 1970.
19. Latour B, Woolgar S, Salk J. *Laboratory Life: The Construction of Scientific Facts*. 2nd edition. Princeton University Press; 1986.
20. Reiss J, Sprenger J. Scientific objectivity. In: Zalta EN, ed. *The Stanford Encyclopedia of Philosophy*. Winter 2020 Edition. Metaphysics Research Lab, Stanford University; 2020 [Accessed November 24, 2020. https://plato.stanford.edu/archives/win2020/entries/scientific-objectivity/].
21. Jasanoff S. *States of Knowledge: The co-Production of Science and Social Order*. Routledge; 2004.
22. Douglas HE. *Science, Policy, and the Value-Free Ideal*. University of Pittsburgh Press; 2009.
23. Bush V. *Science, the Endless Frontier: A Report to the President*. U.S. Government Printing Office; 1945.
24. The President's Scientific Research Board, Steelman JR. *Science and Public Policy: A Program for the Nation. Vol 1*. U.S. Government Printing Office; 1947.
25. National Center for Science and Engineering Statistics (NCSES). *Federal R&D Funding, by Budget Function: Fiscal Years 2019–21*. National Science Foundation; 2021 [Accessed November 20, 2021. https://ncses.nsf.gov/pubs/nsf21315#technical-notes&technical-tables].
26. Gibbons M, Limoges C, Nowotny H, Schwartzman S, Scott P, Trow M. *The New Production of Knowledge: The Dynamics of Science and Research in Contemporary Societies*. Sage Publications; 1994.
27. Guston D. *Between Politics and Science: Assuring the Integrity and Productivity of Research*. Cambridge University Press; 2000.
28. Staff. *Massive 2021 U.S. Spending Bill Leaves Research Advocates Hoping for More*; 2021 [Science. Published December 22, 2020. Accessed November 5, 2021. https://www.science.org/content/article/massive-2021-us-spending-bill-leaves-research-advocates-hoping-more].
29. NCSES. *The State of U.S. Science and Engineering*. National Science Foundation, National Science Board, Science and Engineering Indicators; 2020. Published January 2020 https://ncses.nsf.gov/pubs/nsb20201/u-s-r-d-performance-and-funding#figureCtr886. Accessed 14 May 2021.
30. Brooks H. The scientific advisor. In: Gilpin RG, Wright C, eds. *Scientists and National Policy-Making*. Columbia University Press; 1964:73–96.
31. Zwolenik JJ. Science growth and society: A new perspective. *Pol Sci*. 1971;2(4):457–465.
32. Gault F. Developing a science of innovation policy internationally. In: Lane JI, Fealing KH, Marburger JH, Shipp SS, eds. *The Science of Science Policy*. Stanford University Press; 2011:156–182. https://doi.org/10.1515/9780804781602-011.
33. Stokes DE. *Pasteur's Quadrant: Basic Science and Technological Innovation*. Brookings Institution Press; 1997.

34. Fransman J. Charting a course to an emerging field of "research engagement studies": A conceptual meta-synthesis. *Research for All.* 2018;2(2):185–229. https://doi.org/10.18546/RFA.02.2.02.
35. Minkler M. Linking science and policy through community-based participatory research to study and address health disparities. *Am J Public Health.* 2010;100(S1):S81–S87. https://doi.org/10.2105/AJPH.2009.165720.
36. Lewin K. Action research and minority problems. *J Soc Issues.* 1946;2(4):34–46.
37. Clark F, Illman DL. Dimensions of civic science: Introductory essay. *Forensic Sci Commun.* 2001;23(1):5–27. https://doi.org/10.1177/1075547001023001002.
38. Wandersman A. Community science: Bridging the gap between science and practice with community-centered models. *Am J Community Psychol.* 2003;31(3–4):227–242. https://doi.org/10.1023/A:1023954503247.
39. Funtowicz SO, Ravetz JR. Science for the post-normal age. *Futures.* 1993;25(7):739–755. https://doi.org/10.1016/0016-3287(93)90022-L.
40. Lemos MC, Morehouse BJ. The co-production of science and policy in integrated climate assessments. *Glob Environ Chang.* 2005;15(1):57–68. https://doi.org/10.1016/j.gloenvcha.2004.09.004.
41. Carayannis EG, DFJ C, eds. *Knowledge Creation, Diffusion, and Use in Innovation Networks and Knowledge Clusters: A Comparative Systems Approach across the United States, Europe, and Asia.* Greenwood Publishing Group; 2006.
42. Carayannis EG, von Zedtwitz M. Architecting gloCal (global–local), real-virtual incubator networks (G-RVINs) as catalysts and accelerators of entrepreneurship in transitioning and developing economies: Lessons learned and best practices from current development and business incubation practices. *Dent Tech.* 2005;25(2):95–110. https://doi.org/10.1016/S0166-4972(03)00072-5.
43. Carayannis EG, Campbell DFJ. "Mode 3" and "quadruple helix": Toward a 21st century fractal innovation ecosystem. *IJTM.* 2009;46(3/4):201. https://doi.org/10.1504/IJTM.2009.023374.
44. Nelson A, Kearney W. Science and technology now sit in the center of every policy and social issue. In: *Issues in Science and Technology*; 2021. Published October 19 https://issues.org/science-technology-policy-social-issue-alondra-nelson-interview/. Accessed 5 November 2021.
45. Schwab K. *The Fourth Industrial Revolution.* Currency; 2017.
46. Jasanoff S. *Science and Public Reason.* Routledge; 2012.
47. Jasanoff SS. Contested boundaries in policy-relevant science. *Soc Stud Sci.* 1987;17(2):195–230. https://doi.org/10.1177/030631287017002001.
48. Akerlof K, Tyler C, Foxen SE, et al. A collaboratively derived international research agenda on legislative science advice. *Palgrave Commun.* 2019;5(1):108. https://doi.org/10.1057/s41599-019-0318-6.
49. Tyler C, Akerlof K. Three secrets of survival in science advice. *Nature.* 2019;566(7743):175–177. https://doi.org/10.1038/d41586-019-00518-x.
50. Bimber BA. *The Politics of Expertise in Congress: The Rise and Fall of the Office of Technology Assessment.* State University of New York Press; 1996.
51. Nentwich M. *Parliamentary Technology Assessment Institutions and Practices. A Systematic Comparison of 15 Members of the EPTA Network.* Institut für Technikfolgen-Abschätzung; 2016. https://doi.org/10.1553/ITA-ms-16-02.
52. Bütschi D, Carius R, Decker M, et al. The practice of TA: Science, interaction, and communication. In: Decker M, Ladikas M, Stephan S, Wütscher F, eds. *Bridges between Science, Society and Policy: Technology Assessment — Methods and Impacts.* Berlin, Heidelberg: Wissenschaftsethik und Technikfolgenbeurteilung. Springer; 2004:13–55. https://doi.org/10.1007/978-3-662-06171-8_2.
53. Graves Z, Schuman D. *Science, Technology, & Democracy: Building a Modern Congressional Technology Assessment Office.* Ash Center for Democratic Governance and Innovation,

Harvard Kennedy School; 2020. http://demandprogress.s3.amazonaws.com/reports/Science_Technology_and_Democracy_Building_a_Modern_Congressional_Technology_Assessment_Office.pdf.
54. Guston DH. Boundary organizations in environmental policy and science: An introduction. *Sci Technol Human Values.* 2001;26(4):399–408. https://doi.org/10.1177/016224390102600401.
55. Suhay E, Cloyd ET. Incentives, opportunities, and success? AAAS members' experiences communicating with policymakers. In: *American Geophysical Union (AGU)*; 2018. https://agu.confex.com/agu/fm18/meetingapp.cgi/Paper/394004. Accessed 20 November 2021.
56. Akerlof K.L., Allegra A., Nelson S., Gonnella C., Washbourne C., Tyler C. International perspectives on scientists' roles in legislative policymaking. Pol Sci. n.d. Revise&resubmit.
57. Steel B, Lach D, List P, Shindler B. The role of scientists in the natural resource and environmental policy process: A comparison of Canadian and American publics. *J Environ Syst.* 2001;28(2):133–155.
58. United Nations Statistics Division. *Methodology: Standard Country or Area Codes for Statistical Use (M49).* United Nations: Department of Economic and Social Affairs. Published; 2019. https://unstats.un.org/unsd/methodology/m49/. Accessed 28 July 2019.
59. von Winterfeldt D. Bridging the gap between science and decision making. *PNAS.* 2013;110(Supplement 3):14055–14061. https://doi.org/10.1073/pnas.1213532110.
60. Yoder K. *Why Biden Should Stop Saying 'Listen to the Science.'.* Grist; 2021. Published January 29 https://grist.org/politics/why-is-science-is-so-politicized-blame-the-way-we-talk-about-it/. Accessed 5 November 2021.
61. Slater MD, Rouner D. How message evaluation and source attributes may influence credibility assessment and belief change. *J Mass Commun Quart.* 1996;73(4):974–991. https://doi.org/10.1177/107769909607300415.
62. Mair D, Smillie L, La Placa G, Schwendinger F, Raykovska M, Pásztor Z. *Understanding our Political Nature: How to Put Knowledge and Reason at the Heart of Political Decision-Making.* Luxembourg: Publications Office of the European Union. Published online; 2019.
63. de Vrieze J. Bruno Latour, a veteran of the 'science wars,' has a new mission. *Science.* 2017;358. https://doi.org/10.1126/science.aaq1805. Published online October 10.
64. Ingold K, Gschwend M. Science in policy-making: Neutral experts or strategic policy-makers? *West Eur Polit.* 2014;37(5):993–1018. https://doi.org/10.1080/01402382.2014.920983.
65. Lemos MC, Kirchhoff CJ, Kalafatis SE, Scavia D, Rood RB. Moving climate information off the shelf: Boundary chains and the role of RISAs as adaptive organizations. *Wea Climate Soc.* 2014;6(2):273–285. https://doi.org/10.1175/WCAS-D-13-00044.1.
66. OSTP. *The White House Office of Science and Technology Policy Launches "the Time Is Now: Advancing Equity in Science and Technology Ideation Challenge".* The White House Office of Science and Technology Policy; 2021. Published October 14, 2021. Accessed November 5, 2021 https://www.whitehouse.gov/ostp/news-updates/2021/10/14/the-white-house-office-of-science-and-technology-policy-launches-the-time-is-now-advancing-equity-in-science-and-technology-ideation-challenge/.

CHAPTER THREE

Effects of politicization on the practice of science

Hannah Schmid-Petri[a,*], Nils Bienzeisler[b], and Arista Beseler[a]
[a]University of Passau, Passau, Germany
[b]Karlsruhe Institute of Technology (KIT), Karlsruhe, Germany
*Corresponding author: e-mail address: hannah.schmid-petri@uni-passau.de

Contents

1. Introduction 46
2. Conceptualization and definition of politicization of science 47
3. The politicization of science in the triangular relationship of politics, media, and science and its consequences in the practice of science 49
   3.1 The politicization of science by politics or by political actors 50
   3.2 The politicization of science by (online) media or by journalists 52
   3.3 The politicization of science by science itself or by scientific actors 54
4. Conclusion and practical implications: What can be done to address politicization? 57
References 59

Abstract

Since many of the problems societies face today are complex and, by origin, are scientific (e.g., climate change, the COVID-19 pandemic, etc.), scientific evidence is imperative in many policymaking processes to get a deeper understanding of these issues and possible risks and to derive and justify certain policy measures. The close intertwining of science and politics, however, can have both positive (e.g., growing recognition or reputation, fact-based decision making) and negative consequences (e.g., growing science skepticism, expertocracy, and misuse of scientific credibility to pursue political agendas) for science. The first aim of our paper is to sharpen the theoretical conceptualization of the phenomenon of politicization, and the second aim is to disentangle different drivers (politics and political actors, media and journalists, science and scientists) that may fuel a politicization of science. Based on this, possible effects of politicization for individual scientists and for science as a whole and, thus, for the practice of science are discussed.

Progress in Molecular Biology and Translational Science, Volume 188
ISSN 1877-1173
https://doi.org/10.1016/bs.pmbts.2021.11.005

1. Introduction

Since many of the problems societies face today are complex and, by origin, are scientific (e.g., climate change, the COVID-19 pandemic, etc.), scientific evidence is imperative in many policymaking processes to get a deeper understanding of these issues and possible risks and to derive and justify certain policy measures.[1–3] This close interconnectedness between politics and science is a subject of controversial discussion.

The use of scientific evidence in politics is often associated with the vision that political decisions should be based on the best available facts. Science provides this kind of knowledge, especially as science is normally accorded with a high level of trust and authority.[4–6] The use of scientific "facts" in politics and to bolster political decisions is linked to the hope that science provides neutral, unbiased, and independent facts that can pacify and resolve ideologically colored political controversies.[7–9] However, it is precisely this authority accorded to scientific knowledge that is also used to attack or question certain policies, for example, with the help of contrary scientific evidence or by emphasizing uncertainty (i.e., that the scientific evidence is not yet settled enough to justify specific policy measures).[10,11] Nevertheless, the close connection between science and politics is also viewed critically: Druckman[12] warns of a "crisis of politicization within and beyond science" and highlights the negative consequences that this politicization can bring regarding a growing science skepticism in society. Even though trust in science and/or scientists remains high in most countries,[13] a recently published study reveals that roughly 50% of US adults believe that science is politically motivated or are ambivalent toward the idea of science being political.[14] These beliefs may have severe effects on the credibility of science and are furthermore negatively associated with risk assessments, for example, regarding COVID-19.

Additionally, some researchers criticize the somewhat naive assumption that policy decisions can be derived unambiguously from scientific knowledge.[2,11,15–17] This linear understanding of the relationship between science and politics overlooks the fact that policymaking necessarily involves ideological, social, and cultural value judgments, as policymakers have to weigh up the benefits and costs of different political actions. Accordingly, policies cannot be derived from scientific facts alone but require a debate about value issues.[7,18] If scientific facts are used strategically to settle such value debates instead of informing policymaking best possible, expertocracy might bypass democracy.[4,19]

While the aforementioned issues mostly relate to democratic contexts, the entanglement between science and politics also constitutes an important role in authoritarian systems.[20] As Adelle et al. stated, "the way knowledge is used in policymaking (and therefore also the role of scientists in the creation and use of this knowledge) is related to the type of governance pursued."[21](p64) Non-democratic governments have the capacity to more easily and severely limit their scientific institutions' ability to work independently and thus threaten academic autonomy. For example, in Czechoslovakia, during Soviet times, loyalty to the Communist Party heavily influenced the funding of research equipment or spaces, a person's academic career prospects, and the applied scientific practices.[22] As a result, authoritarian governments can gain direct influence over scientific research projects and, by these means, scientific findings, which might then be deployed to legitimize the regime's actions and decision-making. However, the amount of empirical or theoretical research on the politicization of science in non-democratic states is somewhat limited. Given these circumstances, this chapter will focus on the politicization of science in democratic settings.

These examples show that the close intertwining of science and politics can have both positive (e.g., growing recognition or reputation, fact-based decision making) and negative consequences (e.g., growing science skepticism, expertocracy, and misuse of scientific credibility to pursue political agendas) for science. Moreover, even though there is an increasing debate regarding the phenomenon of politicization and possible consequences for society and science, the concept remains rather fuzzy and is used in different ways in the existing literature; a clear definition or overarching conceptualization is still lacking. Generally, the phenomenon of politicization of science always refers to the interface between science and politics and to the question of how scientific knowledge relates to and is related to political action. Regarding these shortcomings and possible effects that politicization may have on science, the first aim of our paper is to sharpen the theoretical conceptualization of the phenomenon of politicization, and the second aim is to disentangle different drivers that may fuel a politicization of science. Based on this, possible effects of politicization for individual scientists and for science as a whole and, thus, for the practice of science are discussed.

2. Conceptualization and definition of politicization of science

The phenomenon of politicization originates from political science and denotes that a topic is transferred into the realm of politics and/or political debate.[23–27] Thus, in this basic conceptualization, politicization occurs when an issue (e.g., an issue that was originally based in science) is taken up by the political system, when established political actors address it, and/or when they develop initiatives or proposals concerning a specific issue.[26,27] Underlying this conceptualization is an understanding of politicization as a top-down process: politicization occurs when established political actors or institutions decide to tackle certain issues based on strategic considerations or goals and in settings in which the actors have the political, communicative, and organizational opportunities to pursue politicization successfully.[28,29]

However, this kind of conceptualization poses several problems. First, the notion of politics as a clearly delineated sphere does not seem appropriate. Politics and political actions take place everywhere and are not bound to the borders, institutions, arenas, and actors of the established political system.[29–31] Political debates are characterized by various heterogeneous actors who try to make their voice heard and pursue certain political goals, and who do not necessarily have to be part of the political system.[26,27,29] Second, with this macro-perspective on politicization, individual actors with their specific interests and motives cannot be captured theoretically nor empirically. Third, such a conceptualization does not reflect the changed framework conditions brought about by the new communicative possibilities of digital media and the emergence of digital networks. Interestingly, it is precisely these new options of digital communication that make it very easy for actors of all kinds to participate in political debates or even to mobilize for a particular cause.[32]

For these reasons, we propose to understand politics as action in the pursuit of determining certain policies.[30] Thus, any (communicative) action that marks an issue as political or with which political goals are pursued constitutes a political action that drives political processes, expresses political positions, and may finally generate political regulation through the utilization in politics. Therefore, politicization can be conceptualized as a (communicative) act in which something is named as political. This can be understood as a strategic act since calling something political always pursues certain political goals and is often accompanied by the initiation of controversies and debates.[28] The advantages of this conceptualization are that it shifts the perspective to specific processes and actors of politicization with their individual strategies and goals. The approach also incorporates the fact that many actors can be drivers of politicization in many different contexts. In this understanding, the politicization of a particular issue can therefore be initiated by established political actors (top-down) and, for example, by individual civil society actors (bottom-up).

Based on the above-mentioned considerations, the politicization of science thus means the (communicative) act by which science acquires a political meaning. In concrete, the politicization of science can be defined as the process of using scientific knowledge, scientific evidence, or scientific experts to pursue certain political goals. These political goals can be, for example, to better understand a particular scientific topic, to make evidence-based political decisions, to legitimize one's political goals with the help of scientific evidence, to attack one's political opponents, or simply to prevent political regulations by pointing out that scientific uncertainty is still too high. Additionally, the politicization of science also incorporates the fact that the scientific system and the practice of science are influenced by the political settings in which they are embedded. Specific political framework conditions may directly affect how science can be exercised, for example, through funding schemes for specific topics.

The politicization of science can occur in different areas or regarding different facets of politics: regarding certain polities (institutions), regarding politics (actors and political processes), and/or regarding specific policies (instruments).[33] During the COVID-19 pandemic, these connections became particularly clear. Science can show how, for example, vaccinations work and which risks and benefits to the population they offer. However, it is a political question whether the societal benefits of vaccination outweigh individual concerns against it and whether vaccinations should be mandated. In this regard, scientific institutions pleading for compulsory vaccinations or political actors emphasizing scientific findings for or against such a policy are further possible examples. Overall, the politicization of science can always be seen as a strategic (communicative) action to pursue certain political goals. Next, we will elaborate how concretely the politicization of science manifests itself and which drivers may fuel politicization processes. Based on this, we discuss possible consequences the politicization of science can have for the practice of science.

3. The politicization of science in the triangular relationship of politics, media, and science and its consequences in the practice of science

The questions are how the politicization of science can be explained, how exactly it occurs or manifests itself, and who or which groups of actors are the drivers behind the politicization of science. Overall, three groups can be identified as the main drivers behind the politicization of science: politics or political actors, (online) media or journalists, and/or science or scientists.

3.1 The politicization of science by politics or by political actors

As mentioned above, science, scientific evidence, and/or scientific experts play a crucial role in politics, and political actors use science in different ways. We understand political actors very broadly: as all actors pursuing a political goal. This incorporates political actors who are part of the established political system and "outsiders" or "challengers"—for example, single activists.[34] Regarding the different dimensions of politics (see above), the politicization of science probably plays the greatest role in the field of polities and policies.

In these terms, an example of politicization of science in the field of politics could be that specific institutions or advisory boards are founded (and others are not) depending on the political goals certain actors pursue. Another indication of politicization of science is that those boards are staffed with selected scientific experts of whom the position they represent regarding a certain issue is well known.[35]

With regard to the debate about certain policies, different manifestations of politicization processes may occur: the use of science[1] to legitimate and support one's own political position[2] to attack or criticize opponents[2,8,36,37] or[3] to shift political responsibility to the scientific realm.[38–40] Another manifestation of politicization processes occurs when[4] political actors strategically put emphasis on scientific uncertainty, that is, the notion that the scientific evidence regarding a certain issue is not yet settled and that, therefore, more research is needed before policy measures can be taken.[2,10,41,42]

Depending on the issue, the type of actor, and its position in the political system, different strategies are more likely to be used. Those who advocate a particular policy measure attempt to legitimize it on the basis of scientific evidence. Those who oppose that policy measure will either attack the scientific evidence or experts representing the opposing side or bring in other scientific evidence that supports their position or emphasize scientific uncertainty to avoid certain policy regulations. For example, the strategy of instrumentalizing opposing scientific findings to slow down policymaking has been employed by the climate change counter-movement in the US[43,44] and Europe.[45] Such climate change denying or downplaying organizations and think-tanks—which are often implicitly or explicitly linked to conservative, or in some cases far-right, political parties—generate scientific findings meant to undermine the severity of climate change.[44,45] As a result, political entities opposing sustainability policies use this scientific evidence to create an illusion of scientific uncertainty, corroborating their anti-sustainability stance and thus hindering policymaking.[44,45]

One driver behind this politicization of science by political actors is that urgent and complex societal challenges, such as the COVID-19 pandemic, frequently lead to situations in which available scientific knowledge is uncertain while the stakes are high.[1]These complex challenges often touch upon interdisciplinary bodies of knowledge that are characterized by multidimensionality. Scientists respond by competing rather than collaborating in their assessments, thus making the communication more muddled and difficult to understand.[46,47] As of the complex nature of the issues, scientific evidence may suggest contradictory recommendations.[2,9] Accordingly, recent research has observed that scientific facts have been dragged into polarized contemporary political debates.[2,3,32,48] Consequently, science-based policymaking in such fields inevitably leads to a politicization of science by political actors, as they selectively adopt scientific findings to take actions (see above). Thus, certain "chunks of scientific knowledge" acquire a political meaning.

These examples highlight a plethora of possible risks for science as a whole that relate to how single scientists become associated with the politicization of science. The paramount risk is that science or single scientists are seized by politics, and thus, the freedom of scientific research is threatened. A real misuse of science to pursue political goals contributes to a loss of credibility of science and scientists, endangering the societal trust in science and also the reputation of certain scientists. Additionally, the very selective use of scientific evidence to bolster political goals or decisions ("cherry-picking") may insufficiently reflect the state of research on a particular topic, which, in turn, may create a distorted picture of the respective scientific field in public. In the process of politicization, science is drawn into ideological disputes. This violates the postulate of the value-free nature of science and may lead to the public perception that all scientific endeavors are politically motivated (see above). This subsequently reflects negatively on the trustworthiness and authority of science as a whole.

Despite this politicization of science in political debates and negotiations, politics have various expectations of science and furthermore sets its framework conditions.[3,9,49] These—sometimes stronger, sometimes lighter—interventions of politics in science can also be seen as an aspect of the politicization of science. The main aspect with which politics intervenes in science, governs and channels scientific activities are certain incentives and funding schemes. Depending on how these funding schemes are designed, they make certain scientific projects much more attractive than others, which leads to certain topics being more intensively researched than others.

Another sign of the politicization of science by the political sphere is the expectation that science has to deliver solutions to complex problems in a relatively short period (e.g., the development of a vaccine against COVID-19) and, on a more general level, that research has to be useful and can ideally be directly applied to solve concrete societal problems. Linked to these aspects are the demand for scientists to be involved and actively engaged in science communication and policy advice.[9,49,50] These expectations have direct effects on the practice of science. In the meantime, almost every application for third-party funding requires that the submission elaborates on the expected impact of the project for society as well as a section dedicated to planned dissemination activities.

Overall, politics and political actors are among the central drivers behind the politicization of science. This is because modern societies have to tackle some devastating problems in which scientific knowledge plays a central role. In this regard, politicians strategically use scientific evidence (and the authority associated with it) to pursue certain political goals and thus drag science into ideologically colored political debates.

3.2 The politicization of science by (online) media or by journalists

Most people have no direct contact with politics or science, which means that the mass media remain one of the most important sources of information about scientific and political issues.[51] Since the mass media handles a considerable amount of information, reporting is always the result of selection processes in which certain topics or aspects of a topic are chosen and emphasized, and others are left out.[52] In this respect, mass media are an important driver of politicization processes. This is because politicization serves the specific presentation and selection logic of the mass media, and mass media are central gatekeepers for effective politicization since they create publicity. In addition, mass media can mark a scientific issue as a matter of political discussion in their reporting and decide which actors (e.g., scientific versus political actors) they want to cite in their reporting. For example, Chinn et al.[53] showed that the coverage on climate change has become increasingly politicized over the years, meaning that the number of political actors associated with the topic increased over time, whereas the number of scientific voices decreased.

One central characteristic of mass media reporting is personalization, which means that mass media tend to assign causes, consequences, and solutions regarding a certain issue to responsible actors to make a story more vivid and interesting for audiences. Furthermore, personalized reporting transforms a crisis or a certain risk into something tangible, solvable, and thus becomes less threatening or uncertain.[54,55] If individual scientific experts now become visible in public or political debates as part of the politicization of science, this accommodates the logic of the personalization of mass media reporting, which increases the chance that the mass media will report on the issue and thus become a driver of politicization.

Since conflicts always have a high news value for mass media, (supposed) disagreements in science are always interesting to them. Regarding some scientific issues, this may accurately reflect the current discourse within the scientific community. In others, however, it leads to the impression that different groups of scientists disagree when, in fact, this is not the case, or the disagreement is overemphasized in media coverage. This happened, for example, in the reporting on climate change in the past, where journalists created the impression of dueling scientists by contrasting a representative of the scientific consensus with a climate change skeptic in their reporting.[56–58] Additionally, due to the journalistic norm of objectivity, journalists tend to always present both sides of a controversy equally—even if one side may represent only a minority opinion within the scientific community (a phenomenon known as "balance as bias" or "false balance"[52]). The emphasis on conflict and the presentation of dueling scientists creates the public impression that science is still uncertain and not yet settled.[59,60]

The media can, wittingly or unwittingly, influence the audiences' perception of science and scientists depending on their coverage of science and the audiences' media diets.[41,61] For example, depending on whether people mostly use conservatively aligned or liberally aligned media outlets, their trust in science and their perception of scientific issues may vary drastically.[61] Hmielowski et al.[61] concluded that using conservatively aligned media outlets resulted in decreased trust in scientists, while the use of liberal media increased trust in scientists. Thus, the media is imperative in how science and scientists are perceived by the people, which subsequently can affect scientists' role in society.[62]

The new opportunities and changes in scientific work brought about by digitalization also fuel certain politicization processes. Due to the possibilities of digital repositories and digital media, science has become more open and more transparent. This means that the public visibility of science is now extended to knowledge production (open science) and evaluation processes (open peer review). This enables the observation of previously non-transparent areas of the research process. In general, scientific evidence is more easily accessible online, and due to the low access barriers, online scientific findings are accessible at an early stage. This allows the early use of unfinished and preliminary results where there is a greater risk that they will have to be revised and thus open more possibilities for conflicts and discussion.[63] Additionally, the hierarchy of knowledge providers is challenged online—anyone can claim to be an expert, and many heterogeneous actors participate online in public debates.

Overall, online media make scientists more accessible but also more vulnerable to criticism and (personal) attacks. The heterogeneity of actors online and the freedom provided by digital media, which is accompanied by a lack of quality control, bears, in general, many chances for science communication but also poses risks for the reputation and credibility of scientists. Legacy and digital media thus provide a potential forum for politicization processes but are also drivers of politicization, and a certain way of reporting can affect the audience's perception of scientists and science.

3.3 The politicization of science by science itself or by scientific actors

We have shown that the production and application of scientific knowledge in democratic societies are related to different aspects of politicization and driven by factors such as the nature of modern science and recent developments in journalism and the (online) media landscape. However, science is not only politicized by political actors who use science to pursue certain political goals or by the working logic of the media but also by scientists themselves as they participate in public debates and potentially give scientific evidence a political meaning.[3]

In recent years, the need for scientific advice has grown as current crises require urgent decision-making against the background of uncertain knowledge.[9,64] Studies have shown that members of the public appreciate the delegation of political decisions to experts[4,5,16,19] and that politics reinforce the role of scientists as policy advisors.[9] Simultaneously, science is increasingly linked to the mass media and no longer just debated in remote expert circles. The interdependencies have been described as the medialization of science.[49,50] Scientists believe that, with the growing importance of the mass media, it is increasingly relevant for their career and to increase the impact of their research to reach out to the public.[65] Scientists accordingly adapt to the logic of modern mass media and the public and relate to controversies.

Consequently, the politicization of science by scientists might be pushed ahead by the media orientation of scientists and by growing demands of scientific expertise for complex issues directed at scientists.[3]

The subsequent overarching question (and a still ongoing debate) is how actively individual scientists should participate in public debates and how they should position themselves clearly in policy disputes. Pielke[66] identified four typical roles that scientists can take in public and political debates, which are placed on a continuum from "pure scientists" to clear advocates for a particular political position.[67] In his understanding, pure scientists[1] solely focus on their research and do not consider practical applications or the relation of their work to possible societal problems. Such a role model assumes little interaction between science, politics, and society.[66 (p15)] This role is hardly relevant in today's world, where scientific experts participate at multiple and very different levels of society. Science arbiters[2] or the decisionist model assume that politicians define the current problems and that decisions are made in the political system. To solve these societal challenges defined by policy makers, science should then conduct research and provide the political sphere with the best solutions. These will then be implemented by political decision-makers. Thus, the science arbiter remains removed from explicit considerations of policy and politics but, unlike pure scientists, informs policymaking through applied research and assessments when required.[66 (p16)] The implicit assumption underlying this role conception is that concrete scientific activities are governed by politics. Beyond that, there is little dialogue with society. The honest broker or pragmatic model[3] stands for scientists who choose to participate in the political process (e.g., as experts on advisory boards) but do not pursue a certain political agenda.[66 (pp17–19)] Following the norm of impartiality, they present different policy alternatives to decision-makers while taking stakeholders' concerns into account. Such an understanding of the interaction between science, politics, and society rests on the so-called knowledge-broking strategies.[68] The goal of these strategies is an exchange of scientists and policymakers to maximize the scope of action for policymaking by presenting a selection of options. This role assumes that both spheres influence each other. Scientists siding with a societal or political group and promoting a certain political agenda are defined as issue advocates.[4,66 (pp15–16)] Thus, they affirm an expertocratic understanding of the relationship of policymaking rooted in a linear understanding of the relationship between science and politics.

How the politicization of science by scientists plays out depends on either of the last two roles that scientists assume.[11] Acting as honest brokers, scientists may use their credibility to point toward societal issues or possible policies. In these terms, the politicization of scientific findings may foster deliberation and democratic discourse, presenting a portfolio of policy choices and tradeoffs. In contrast, scientists speaking out as issue advocates can bias policymaking in the same way as the political actors they thereby become. An example could be that scientists use their credibility to legitimize and support their political position, as revealed by surveys among scientists in nanotechnology[69] or among climate scientists.[48,70] This bears the risk that scientists selectively highlight policy options motivated by their own political preferences.[3,11] Thus, scientists become political actors. Although the efforts pursued by these scientists can be tremendously important, the presentation of findings to promote only a distinctive subset of policy options in public was criticized for distorting debates.[11] Accordingly, issue advocates are especially prone to politicize science. Furthermore, scientists could raise several strains for democratic societies in cases where scientists only pretend to take value-free positions but strategically use their expertise and reputation to advocate for certain policy goals and thus are not transparent concerning their role and aims.[66] When individual scientists stretch their role in this way, this may harm the image, credibility, and reputation of science as a whole.[11,17] Moreover, when scientists behave like political actors, they are attacked in discourse just as harshly as genuine political actors, which further damages scientific credibility and reputation.

In general, it can be said that it is difficult for individual scientists to participate in public debates that are ideologically driven. In these settings, a "neutral" and fact-based communication of information is hardly possible: political actors argue with scientific arguments, and experts are perceived as political actors, even if they might not see themselves that way.

Overall, we see that the demand for scientific advice has increased. As a result, scientists are increasingly expected to speak out in public debates. For an informed debate on the complex challenges of today, science can make valuable contributions; thus, it is vital that scientists feed their knowledge and scientific evidence into the political process. For science, however, this poses the risk that value-free science will be drawn into ideological debates and instrumentalized there. This phenomenon, in turn, could have the consequence in the long run that the practice of science is influenced by certain worldviews in the sense that scientific knowledge is generated regarding its consequences or that certain topics are not researched because they are not politically in vogue.[47]

4. Conclusion and practical implications: What can be done to address politicization?

The politicization of science raises several challenges to the scientific community, epitomizing the need for critical reflection on contemporary science communication and the interface between science and politics. As we have shown, there are close links between science and politics—an interconnection that can be very fruitful, but that can also pose several challenges for science as a whole as well as for individual scientists. In the following, we discuss what actors from politics, media, and academia can do to address politicization.

First, political actors use scientific evidence to pursue certain political goals. This could be the comparatively "neutral" goal to better understand the scientific background of an issue or the consequences certain policy regulations may have. Moreover, political actors use scientific evidence strategically (and selectively) in ideological debates to legitimate their political positions, to criticize opponents, or to avoid binding policy regulations. Since politics depends on scientific knowledge, there will always be a politicization of science within politics. Thus, the use of scientific knowledge is an inherent (and useful) characteristic of the political process. However, political actors should, of course, not intentionally distort scientific evidence, instrumentalize experts, or attack scientists to bolster certain political goals.

Second, as described above, mass media often fuel politicization, as they emphasize conflict and disagreement and tend to personalize stories. To avoid further politicization and to pacify public debates, mass media should ideally provide a content-oriented discourse based on facts. An example of this is the reporting on the CRISPR/Cas technology. An analysis of the relating press coverage in North America revealed a predominant topical focus on human health, with most of the articles defining the method as a "gene-editing tool" and being either positive (38%), mostly positive (27%), or neutral (29%) in tone. Moreover, 96% of the analyzed articles portrayed the CRISPR/Cas technology as beneficial concerning health topics, giving the appearance of strong scientific unity.[71] This approach is followed by the concept of "constructive journalism"[72] which tries to foster solution-oriented reporting instead of negative and conflict-based news. In this context, qualitatively high science journalism is indispensable, as it serves as the key mediator between science and other societal groups.

Third, scientists inevitably become part of political disputes when they speak out for certain policies in public, whether they like it or not. Even when scientists are drawn into political disputes as experts, they operate in a public arena that is determined by factors outside of science and constrained by values such as economic concerns or moral codes. If we recall the role that scientific facts played in the debate over mandatory vaccinations, we see that there are reasons for epidemiologists to speak out for such policies. However, this communicative act gives scientific findings political significance and thus leads to a politicization of science. At this juncture, science has to compete with religious concerns, for example, as the public arena is characterized by competing justifications and different legitimations.[47] Accordingly, scientists need to make their standpoints and value positions explicit if they engage in partisan disputes.[3,11,47,67] Transparency and accuracy are important to build and maintain trust in science and scientists in public.

That being said, each scientist has to decide which role they want to play in political discussions—if they want to provide the range of scientific results or if they want to engage as an activist. In the role of an activist, scientists can be drawn into mere ideological debates and thus leave the common ground of science. Therefore, each scientist may adopt knowledge broker strategies that explicitly build on such considerations.[68,73] Yet, this task may exceed the capabilities of most scientists. Previous research accordingly highlighted the need for science-policy intermediaries, a separate kind of expert organization that engages with both science and policy, facilitating the flow of information between science and society.[74,75] Such scientific organizations could disseminate the aggregated state of the art of science to society and politics and thereby contribute to close science–public divides. Another advantage of an institutionalized aggregation of state of the art in a specific research field would be that it "protects" the single scientist as they do not have to expose themselves in public and therefore become less vulnerable to attacks. Moreover, it is important to note that scientific organizations cannot fulfill the role science journalism has in democratic societies. With their science communication, these organizations always pursue strategic goals (e.g., to increase the visibility and reputation of their institution) and thus implicitly drive politicization.

In conclusion, it can be stated that the nature of modern science and societal challenges requires interaction between science and society, as science is no longer confined to the walls of research institutions. On the one hand, tensions between scientific knowledge, technical feasibility, and ethical justifiability are becoming more apparent in modern democratic societies.[2,76]

Normative and ethical issues associated with science are no longer the responsibility of a small audience of scientists but relocated and negotiated within the public sphere.[3,47,77] On the other hand, many items discussed on the political agenda today are actually based on observations in science. Thus, the contribution of science to policymaking is indispensable.[9] As a result, scholars such as Jotterand,[47] Lupia,[37] and Scheufele[3] emphasized the importance of proactively integrating bench science and social science. For science communicators, including molecular biologists and clinical physicians, it is vital to develop more realistic expectations about audiences, to critically reflect on the normative and ethical implications of scientific research, and to close divides between science and the public. Researchers also need to be aware of the limits of their profession and the temptations of strategic political communication, as there is no scientific answer to value disputes, like the disputes about mandatory vaccinations during the global pandemic.

References

1. Funtowicz SO, Ravetz JR. Uncertainty, complexity and post-normal science. *Environ Toxicol Chem*. 1994;13(12):1881–1885. https://doi.org/10.1002/etc.5620131203.
2. Sarewitz D. How science makes environmental controversies worse. *Environ Sci Policy*. 2004;7(5):385–403. https://doi.org/10.1016/j.envsci.2004.06.001.
3. Scheufele DA. Science communication as political communication. *Proc Natl Acad Sci U S A*. 2014;111(Suppl 4):13585–13592. https://doi.org/10.1073/pnas.1317516111.
4. Brossard D, Shanahan J. Do citizens want to have their say? Media, agricultural biotechnology, and authoritarian views of democratic processes in science. *Mass Commun Soc*. 2003;6(3):291–312. https://doi.org/10.1207/S15327825MCS0603_4.
5. Howell EL, Wirz CD, Scheufele DA, Brossard D, Xenos MA. Deference and decision-making in science and society: How deference to scientific authority goes beyond confidence in science and scientists to become authoritarianism. *Public Underst Sci*. 2020;1–19. https://doi.org/10.1177/0963662520962741.
6. Schäfer MS. Mediated trust in science: Concept, measurement and perspectives for the 'science of science communication'. *JCOM*. 2016;15(05). https://doi.org/10.22323/2.15050302.
7. Dietz T. Bringing values and deliberation to science communication. *Proc Natl Acad Sci U S A*. 2013;110(Suppl 3):14081–14087. https://doi.org/10.1073/pnas.1212740110.
8. Jasanoff S. Contested boundaries in policy-relevant science. *Soc Stud Sci*. 1987;17(2):195–230. https://doi.org/10.1177/030631287017002001.
9. Weingart P. Scientific expertise and political accountability: paradoxes of science in politics. *Sci Public Policy*. 1999;26(3):151–161. https://doi.org/10.3152/147154399781782437.
10. Bolsen T, Druckman JN. Counteracting the politicization of science. *J Commun*. 2015;65(5):745–769. https://doi.org/10.1111/jcom.12171.
11. Pielke RA. When scientists politicize science: Making sense of controversy over the skeptical environmentalist. *Environ Sci Policy*. 2004;7(5):405–417. https://doi.org/10.1016/j.envsci.2004.06.004.
12. Druckman JN. The crisis of politicization within and beyond science. *Nat Hum Behav*. 2017;1(9):615–617. https://doi.org/10.1038/s41562-017-0183-5.

13. Funk C, Tyson A, Kennedy B, Johnson C. Science and scientists held in high esteem across global publics. *Published September.* 2020;29. https://www.pewresearch.org/science/2020/09/29/science-and-scientists-held-in-high-esteem-across-global-publics/. Accessed 30 September 2021.
14. McLaughlin DM, Mewhirter J, Sanders R. The belief that politics drive scientific research & its impact on COVID-19 risk assessment. *PLoS One.* 2021;16(4):e0249937. https://doi.org/10.1371/journal.pone.0249937.
15. Grundmann R, Rödder S. Sociological perspectives on earth system modeling. *J Adv Model Earth Syst.* 2019;11(12):3878–3892. https://doi.org/10.1029/2019MS001687.
16. Post S, Bienzeisler N, Lohöfener M. A desire for authoritative science? How citizens' informational needs and epistemic beliefs shaped their views of science, news, and policymaking in the COVID-19 pandemic. *Public Underst Sci.* 2021;1–19. https://doi.org/10.1177/09636625211005334.
17. Lackey RT. Science, scientists, and policy advocacy. *Conserv Biol.* 2007;21(1):12–17. https://doi.org/10.1111/j.1523-1739.2006.00639.x.
18. Weber M. Die Objektivität sozialwissenschaftlicher und sozialpolitischer Erkenntnis. *Archiv für Sozialwissenschaft und Sozialpolitik.* 1904;19(1):22–87. https://nbn-resolving.org/urn:nbn:de:0168-ssoar-50770-8.
19. Hibbing JR, Theiss-Morse E. *Stealth Democracy: Americans' Beliefs about how Government Should Work.* Cambridge Univ. Press; 2002.
20. Zhang JY. The 'credibility paradox' in China's science communication: Views from scientific practitioners. *Public Underst Sci.* 2015;24(8):913–927. https://doi.org/10.1177/0963662515598249.
21. Adelle C, Pereira L, Görgens T, Losch B. Making sense together: The role of scientists in the coproduction of knowledge for policy making. *Science and Public Policy.* 2020;47(1):56–66. https://doi.org/10.1093/scipol/scz046.
22. Linkova M, Stockelova T. Public accountability and the politicization of science: The peculiar journey of Czech research assessment. *Science and Public Policy.* 2012;39(5):618–629. https://doi.org/10.1093/scipol/scs039.
23. Buzan B, Waever O, Wilde Jd. *Security: A New Framework for Analysis.* 12th ed. Lynne Rienner; 2013.
24. Wilde Jd. No polity for old politics? A framework for analyzing the politicization of european integration. *J Eur Integr.* 2011;33(5):559–575. https://doi.org/10.1080/07036337.2010.546849.
25. Wilde Jd, Zürn M. Can the politicization of European integration be reversed? *Journal of Common Market Studies.* 2012;50:137–153. https://doi.org/10.1111/j.1468-5965.2011.02232.x.
26. Zürn M. The politicization of world politics and its effects: Eight propositions. *Eur Pol Sci Rev.* 2014;6(1):47–71. https://doi.org/10.1017/S1755773912000276.
27. Zürn M, Binder M, Ecker-Ehrhardt M. International authority and its politicization. *Int Theory.* 2012;4(1):69–106. https://doi.org/10.1017/S1752971912000012.
28. Wiesner C. Politicisation, politics and democracy. In: Wiesner C, ed. *Rethinking Politicisation in Politics, Sociology and International Relations.* Palgrave Macmillan; 2021:19–42.
29. Palonen K, Wiesner C, Selk V, et al. Rethinking politicisation. *Contemp Polit Theory.* 2019;18(2):248–281. https://doi.org/10.1057/s41296-019-00326-y.
30. Palonen K. Four times of politics: Policy, polity, politicking, and politicization. *Alternatives.* 2003;28(2):171–186. https://doi.org/10.1177/030437540302800202.
31. Selk V. Three concepts of politicisation: Republican, deliberative, and agonistic. In: Wiesner C, ed. *Rethinking Politicisation in Politics, Sociology and International Relations.* Palgrave Macmillan; 2021:43–62.

32. Bennett WL, Pfetsch B. Rethinking political communication in a time of disrupted public spheres. *J Commun*. 2018;68(2):243–253. https://doi.org/10.1093/joc/jqx017.
33. Treib O, Bähr H, Falkner G. Modes of governance: Towards a conceptual clarification. *J Eur Publ Policy*. 2007;14(1):1–20. https://doi.org/10.1080/13501760601071406.
34. Kriesi H. The populist challenge. *West Eur Polit*. 2014;37(2):361–378. https://doi.org/10.1080/01402382.2014.887879.
35. McGee S. Ideology and politicization in public bioethics. *Camb Q Healthc Ethics*. 2011;20(1):73–84. https://doi.org/10.1017/S0963180110000630.
36. Freudenburg WR, Gramling R, Davidson DJ. Scientific certainty argumentation methods (SCAMs): Science and the politics of doubt. *Sociological Inquiry*. 2008;78(1):2–38. https://doi.org/10.1111/j.1475-682X.2008.00219.x.
37. Lupia A. Communicating science in politicized environments. *Proc Natl Acad Sci U S A*. 2013;110(Suppl 3):14048–14054. https://doi.org/10.1073/pnas.1212726110.
38. Pepermans Y, Maeseele P. Democratic debate and mediated discourses on climate change: From consensus to de/politicization. *Environ Commun*. 2014;8(2):216–232. https://doi.org/10.1080/17524032.2014.906482.
39. Maeseele P. On news media and democratic debate: Framing agricultural biotechnology in Northern Belgium. *International Communication Gazette*. 2011;73(1–2):83–105. https://doi.org/10.1177/1748048510386743.
40. Ylönen M, Litmanen T, Kojo M, Lindell P. The (de)politicisation of nuclear power: The Finnish discussion after Fukushima. *Public Underst Sci*. 2017;26(3):260–274. https://doi.org/10.1177/0963662515613678.
41. Bolsen T, Druckman JN, Cook FL. How frames can undermine support for scientific adaptations: Politicization and the status-quo bias. *Public Opin Q*. 2014;78(1):1–26. https://doi.org/10.1093/POQ/NFT044.
42. Schmid-Petri H. Politicization of science: How climate change skeptics use experts and scientific evidence in their online communication. *Clim Change*. 2017;145(3–4):523–537. https://doi.org/10.1007/s10584-017-2112-z.
43. Brulle RJ. Denialism: Organized Opposition to Climate Change Action in the United States. In: Konisky DM, ed. *Handbook of U.S. Environmental Policy*. In: Edward Elgar Publishing; 2020.
44. Brulle RJ. Institutionalizing delay: Foundation funding and the creation of U.S. climate change counter-movement organizations. *Clim Change*. 2014;122(4):681–694. https://doi.org/10.1007/s10584-013-1018-7.
45. Almiron N, Boykoff M, Narberhaus M, Heras F. Dominant counter-frames in influential climate contrarian European think tanks. *Clim Change*. 2020;162(4):2003–2020. https://doi.org/10.1007/s10584-020-02820-4.
46. Gibbons M. *The New Production of Knowledge: The Dynamics of Science and Research in Contemporary Societies*. Sage; 1994.
47. Jotterand F. The politicization of science and technology: Its implications for nanotechnology. *J Law Med Ethics*. 2006;34(4):658–666. https://doi.org/10.1111/j.1748-720X.2006.00084.x.
48. Post S. Communicating science in public controversies: Strategic considerations of the German climate scientists. *Public Underst Sci*. 2016;25(1):61–70. https://doi.org/10.1177/0963662514521542.
49. Weingart P. Science and the media. *Research Policy*. 1998;27(8):869–879. https://doi.org/10.1016/S0048-7333(98)00096-1.
50. Rödder S. Reassessing the concept of a medialization of science: A story from the "book of life". *Public Underst Sci*. 2009;18(4):452–463. https://doi.org/10.1177/0963662507081168.
51. Maier M, Rothmund T, Retzbach A, Otto L, Besley JC. Informal learning through science media usage. *Educational Psychologist*. 2014;49(2):86–103. https://doi.org/10.1080/00461520.2014.916215.

52. Bennett WL. *News: The Politics of Illusion*. 10th ed. The University of Chicago Press; 2016.
53. Chinn S, Hart PS, Soroka S. Politicization and polarization in climate change news content, 1985–2017. *Science Communication*. 2020;42(1):112–129. https://doi.org/10.1177/1075547019900290.
54. Adam S, Maier M. Personalization of politics a critical review and agenda for research. *Ann Int Commun Assoc*. 2010;34(1):213–257. https://doi.org/10.1080/23808985.2010.11679101.
55. van Aelst P, Sheafer T, Hubé N, Vreese Cd. Personalization. In: Esser F, Hopmann DN, eds. *Comparing Political Journalism*. Routledge; 2016:130–148.
56. Dearing JW. Newspaper coverage of maverick science: Creating controversy through balancing. *Public Underst Sci*. 1995;4(4):341–361. https://doi.org/10.1088/0963-6625/4/4/002.
57. Ruhrmann G, Guenther L, Kessler SH, Milde J. Frames of scientific evidence: How journalists represent the (un)certainty of molecular medicine in science television programs. *Public Underst Sci*. 2015;24(6):681–696. https://doi.org/10.1177/0963662513510643.
58. Stocking SH, Holstein LW. Manufacturing doubt: journalists' roles and the construction of ignorance in a scientific controversy. *Public Underst Sci*. 2009;18(1):23–42. https://doi.org/10.1177/0963662507079373.
59. Dunwoody S, Peters HP. Mass media coverage of technological and environmental risks: A survey of research in the United States and Germany. *Public Underst Sci*. 1992;1(2):199–230. https://doi.org/10.1088/0963-6625/1/2/004.
60. Stocking SH. How journalists deal with scientific uncertainty. In: Friedman SM, Dunwoody S, Rogers CL, eds. *Communicating Uncertainty: Media Coverage of New and Controversial Science*. Routledge; 1999:23–42.
61. Hmielowski JD, Feldman L, Myers TA, Leiserowitz A, Maibach E. An attack on science? Media use, trust in scientists, and perceptions of global warming. *Public Underst Sci*. 2014;23(7):866–883. https://doi.org/10.1177/0963662513480091.
62. Fowler EF, Gollust SE. The content and effect of politicized health controversies. *Ann Am Acad Pol Soc Sci*. 2015;658(1):155–171. https://doi.org/10.1177/0002716214555505.
63. Neuberger C, Weingart P, Fähnrich B, et al. Der Digitale Wandel Der Wissenschaftskommunikation. In: *Wissenschaftspolitik im Dialog*. Eine Schriftenreihe der Berlin-Brandenburgischen Akademie der Wissenschaften; 2021:16.
64. Brüggemann M, Lörcher I, Walter S. Post-normal science communication: Exploring the blurring boundaries of science and journalism. *Journal of Science Communication*. 2020;19(3). https://doi.org/10.22323/2.19030202.
65. Peters HP. Gap between science and media revisited: Scientists as public communicators. *Proc Natl Acad Sci U S A*. 2013;110(Suppl 3):14102–14109. https://doi.org/10.1073/pnas.1212745110.
66. Pielke RA. *The Honest Broker: Making Sense of Science in Policy and Politics*. Cambridge University Press; 2007.
67. Schrögel P, Humm C. Science communication, advising, and advocacy in public debates. In: Leßmöllmann A, Dascal M, Gloning T, eds. *Science Communication*. De Gruyter; 2019:485–514.
68. Meyer M. The rise of the knowledge broker. *Science Communication*. 2010;32(1):118–127. https://doi.org/10.1177/1075547009359797.
69. Corley EA, Scheufele DA, Hu Q. Of risks and regulations: How leading U.S. nanoscientists form policy stances about nanotechnology. *J Nanopart Res*. 2009;11(7):1573–1585. https://doi.org/10.1007/s11051-009-9671-5.
70. Post S, Ramirez N. Politicized science communication: Predicting scientists' acceptance of overstatements by their knowledge certainty, media perceptions, and presumed media effects. *Journalism & Mass Communication Quarterly*. 2018;95(4):1150–1170. https://doi.org/10.1177/1077699018769668.

71. Marcon A, Master Z, Ravitsky V, Caulfield T. CRISPR in the North American popular press. *Genet Med*. 2019;21(10):2184–2189. https://doi.org/10.1038/s41436-019-0482-5.
72. Mast J, Coesemans R, Temmerman M. Constructive journalism: Concepts, practices, and discourses. *Journalism*. 2019;20(4):492–503. https://doi.org/10.1177/1464884918770885.
73. Duncan R, Robson-Williams M, Edwards S. A close examination of the role and needed expertise of brokers in bridging and building science policy boundaries in environmental decision making. *Palgrave Commun*. 2020;6(1). https://doi.org/10.1057/s41599-020-0448-x.
74. Guston DH. Boundary organizations in environmental policy and science: An introduction. *Science, Technology, & Human Values*. 2001;26(4):399–408. https://doi.org/10.1177/016224390102600401.
75. Bednarek AT, Shouse B, Hudson CG, Goldburg R. Science-policy intermediaries from a practitioner's perspective: The Lenfest Ocean program experience. *Science and Public Policy*. 2016;43(2):291–300. https://doi.org/10.1093/scipol/scv008.
76. Iyengar S, Massey DS. Scientific communication in a post-truth society. *Proc Natl Acad Sci U S A*. 2019;116(16):7656–7661. https://doi.org/10.1073/pnas.1805868115.
77. Wirz CD, Scheufele DA, Brossard D. Societal debates about emerging genetic technologies: Toward a science of public engagement. *Environ Commun*. 2020;14(7):859–864. https://doi.org/10.1080/17524032.2020.1811478.

CHAPTER FOUR

Politics v. science: How President Trump's war on science impacted public health and environmental regulation

Romany M. Webb[a,*] and Lauren Kurtz[b]
[a]Sabin Center for Climate Change Law, New York, NY, United States
[b]Climate Science Legal Defense Fund, New York, NY, United States
*Corresponding author: e-mail address: rmw2149@columbia.edu

Contents

1. The Silencing Science Tracker 66
2. Anti-science actions under Trump 67
   2.1 Censorship and self-censorship 67
   2.2 Personnel changes 69
   2.3 Budget cuts 70
   2.4 Research hindrance 71
   2.5 Bias and misrepresentation 72
   2.6 Agencies affected 74
3. The Biden administration's approach to science 74
4. Where to from here? Restoring the role of science 76
References 77

Abstract

During his campaign for president, Joe Biden vowed to "end the politics and follow the science" when dealing with the COVID-19 pandemic and other public health and environmental crises. He was immediately criticized by then President Trump, who cast "listen[ing] to the scientists" as something only a fool would do, and warned that it would result in a "massive [economic] depression." It is hardly surprising that Trump would take that position. After all, the Trump administration routinely prioritized economic interests, and worked tirelessly to remove what it viewed as unnecessary regulatory burdens on economic activity. The Trump administration regularly suppressed, downplayed, or simply ignored scientific research demonstrating the need for regulation to protect public health and the environment. The Biden administration has vowed to reverse course, but faces challenges in doing so due to the widespread assault on science led by former President Trump.

The Trump administration's efforts to undermine science are documented in the Silencing Science Tracker, an online database, which records anti-science actions taken

Progress in Molecular Biology and Translational Science, Volume 188
ISSN 1877-1173
https://doi.org/10.1016/bs.pmbts.2021.11.006

by the federal, state, and local governments. Drawing on more than four years of tracker data—from Trump's election to Biden's inauguration—we show that the Trump presidency fundamentally changed how federal government agencies perform, use, and communicate scientific research. While the Biden administration has taken initial steps to undo some of those changes, it still has significant work to do to restore the role of science in federal government decision-making. Its task is made more difficult by the public distrust of science engendered by the Trump presidency.

1. The Silencing Science Tracker

The Silencing Science Tracker is a joint project of the Climate Science Legal Defense Fund[1] and Columbia Law School's Sabin Center for Climate Change Law.[2] The tracker records reported attempts by federal, state, and local government actors to restrict or prohibit scientific research, education, or discussion, or the publication or use of scientific information ("anti-science actions"). According to the tracker, 346 anti-science actions were taken by the federal government between President Trump's election and President Biden's inauguration (i.e., from November 8, 2016 to January 20, 2021). During the same period, a further 156 anti-science actions were taken by state and local governments, but those actions are not discussed in this chapter.

Federal actions recorded in the tracker are categorized as follows:

1. government censorship;
2. self-censorship;
3. budget cuts;
4. personnel changes;
5. research hindrance; and
6. bias and misrepresentation.

Within the above categories, the tracker records actions taken by the federal executive and Congress, except legislative proposals. Many tracker entries involve multiple types of action or actors. For the purposes of this analysis, those entries were separated into their component parts, resulting in 428 unique instances of anti-science behavior, each of which involves one type of action (i.e., from the list above), performed by one actor (e.g., a specific executive agency). The figures shown below were calculated based on that total.. There is reason to believe that many anti-science actions were not reported and thus are not captured in the tracker, and therefore the total represents a conservative estimate of anti-science actions taken between November 2016 and January 2021. In a survey conducted by the Office of Inspector General for the Environmental Protection Agency (EPA), nearly 400 EPA scientists said they had observed violations of the agency's scientific integrity policy in the second half of 2018, but did not report them due to "fear of retaliation, belief that reporting would make no difference, perceived suppression or interference by Agency leadership, and belief that politics and policy outweigh science."[3] Given the Trump administration's widespread and continued attacks on science, similar concerns were likely also held by scientists at other federal agencies throughout the second half of the Trump presidency. It is, therefore, reasonable to assume that a large number of anti-science actions went unreported.

2. Anti-science actions under Trump

During his first presidential campaign, Donald Trump promised to "ensure... total [scientific] transparency and accountability without political bias."[4] That was a promise he didn't keep. As detailed further below, during his four years in office, former President Trump led a concerted effort to undermine federal scientific research, particularly research relating to climate change. The Trump administration's attacks on climate science dovetailed neatly with one of the former President's key goals: to roll-back climate regulations that scientific research shows would advance public health and environmental quality. Faced with this contradiction, the Trump administration sought to restrict access to scientific information or cast doubt on its veracity, thereby limiting public understanding of the issues and reducing possible opposition to the administration's plans. Those actions created a culture of fear among federal scientists, leading some to voluntarily suppress or distort information at odds with former President Trump's agenda. Many of the scientists who did speak out were removed from their positions, while others were prevented from conducting further research on topics deemed "controversial," such as climate change.

2.1 Censorship and self-censorship

During President Trump's time in office (including the transition period), there were 154 documented instances of federal government censorship of scientists, and 19 instances of scientists engaging in self-censorship. Approximately 72% involved the suppression of information about climate change. This began even before President Trump took office. In November 2016, staff at Centers for Disease Control and Prevention (CDC) deleted content discussing the relationship between climate change and human health from at least four webpages, reportedly to "avoid drawing the new president's ire." Similar changes were made to other federal agency websites after President Trump took office. In total, during the Trump administration, climate change and other scientific information was removed from the websites of twelve federal bodies, in most cases at the direction of administration officials.[5] This made it more difficult for Americans to educate themselves about climate change and other scientific issues, which may, in turn, have made it easier for the Trump administration to act on those issues by allowing them to "fly under the radar" or obscuring the consequences of administration action.

The Trump administration also removed scientific information from regulatory documents. For example, in or around August 2018, administration officials deleted information[6] on the local health effects of climate change from regulatory documents supporting the weakening of greenhouse gas emissions controls. Again, this helped the Trump administration advance its deregulatory agenda, including by casting doubt on the need for climate regulations. Trump administration officials also attempted to suppress information that could lead to demands for stricter regulation (e.g., because it shed additional light on the impacts of climate change or demonstrated the inadequacy of that existing attempts to address it).[7] This could have lasting consequences, making it more difficult for the current and future administrations to take regulatory action, due to a lack of information or sense of urgency.

This type of scientific censorship was widespread during the Trump administration, having been documented at 20 federal bodies—more than any other type of anti-science action. Notably however, the number of documented instances of government censorship declined slightly over time, falling by 26% from 2017 to 2018, a further 18% in 2019, and 10% more in 2020. This is not necessarily good news; it may simply reflect the fact that less science was done because of personnel changes, budget cuts, and other anti-science actions taken by the Trump Administration. There is also reason to believe that the attacks on science created a culture of fear among federal employees and led some to self-censor. A survey conducted in 2016—before President Trump's election—found that 72% of EPA scientists felt they could "openly express scientific opinions about the Agency's scientific work without fear of retaliation."[8] That number dropped to just 57% in a repeat survey conducted in 2018—almost two years into Trump presidency.[8] In the 2018 survey, over 600 scientists said their "management chains do not consistently stand behind scientific staff who put forth scientifically defensible positions, including those that may be controversial."[8] It is, then, hardly surprising that some scientists would choose to self-censor. However, while understandable, such behavior could undermine public trust in science by creating the impression that scientists "pick and choose" what to disclose and regularly "hide" information. Both self- and government censorship may also cause the public to question whether research conducted or overseen by federal scientist is truly impartial and lead some to belief that such research is inherently political and thus untrustworthy.

2.2 Personnel changes

In addition to suppressing information, the Trump administration also sought to restrict or prevent further climate change research, including by removing[9] and reassigning[10] federal government scientists. This reduced the capacity of key science agencies. For example, the U.S. Geological Survey—the science arm of the U.S. Department of the Interior (DOI)—lost 150 staff scientists or over 2% of its total scientific workforce between 2016 and 2020.[11] During the same period, 672 scientists left the Environmental Protection Agency (EPA), resulting in 6% decline its total scientific workforce.[12]

As well as reducing federal agencies' internal scientific expertise, the Trump administration also sought to limit their access to outside experts. To that end, in June 2019, President Trump issued an Executive Order directing each federal agency to eliminate at least one-third of its current scientific advisory committees.[13] Following the order, at least nine advisory committees across the Department of Commerce, Department of Energy,[14] DOI,[15] and EPA[16] were terminated. Many other advisory committees (e.g., at EPA,[17] DOI, and the Department of Labor[18]) were unofficially suspended or had their membership changed, with independent scientists replaced by industry representatives.[19] In some cases, the new appointees lacked appropriate expertise, leading to concerns that the Trump administration was stacking advisory committees with favored "experts" who would be unable or unwilling to question the science behind its decisions. This was, perhaps, most obvious at EPA where members of the committee responsible for advising on particulate matter pollution themselves warned that they did "not have the breadth and depth of knowledge or expertise . . . necessary to adequately advise the EPA and to meet the statutory requirement for a thorough and accurate review" of existing or proposed particulate matter controls.[20]

The dismantling of science advisory committees furthered the Trump administration's goal of rolling back climate change regulations in several ways. Perhaps most importantly, it limited external review of the scientific bases for the Trump administration's deregulatory actions, many of which were subsequently struck down by the courts on the basis that they were not supported by the available evidence or the result of reasoned decision-making. Expert review could have identified those flaws before action was taken, but the Trump administration seemingly thought it was more important to move ahead quickly and avoid the possibility of anyone questioning its approach. The administration's actions also had the effect of restricting federal agency and therefore public access to information that might justify action on climate change. That appears to have been the Trump administration's goal when it disbanded the Advisory Committee for the Sustained National Climate Assessment while it was in the process of drafting a report intended to assist government officials to use the National Climate Assessment in long-term planning.[21] These types of actions again undermine trust in science by suggesting that scientific research and findings can be easily manipulated to achieve political ends.

2.3 Budget cuts

Under President Trump, federal agencies also faced pressure to reduce spending on scientific research, with the administration proposing deep across-the-board cuts in many budget cycles.[22] Those proposals were largely rejected by Congress, which actually increased research funding during the Trump presidency.[22] Nevertheless, many existing research programs had their funding cut or entirely eliminated. For example, in August 2017, DOI halted ongoing research into the health impacts of mountaintop removal coal mining, pending a review.[23] The research was officially cancelled in April 2018, with DOI claiming that it was "redundant."[23] A subsequent investigation found that DOI's then principal deputy assistant secretary for land and minerals management, Katherine MacGregor, had pushed for cancellation of the study after repeatedly meeting with the National Mining Association and companies engaged in mountaintop removal coal mining.[23] Another DOI official, Landon "Tucker" Davis, reportedly said the study should be cancelled because "science was a Democrat thing," reinforcing the idea that scientific research is inherently political.[23]

Further supporting this view, under President Trump, some federal agencies also began requiring new research programs to be reviewed by political appointees to ensure they "promote the [Trump administration's] priorities."[24] It appears that appointees often used the review process to further deregulatory initiatives, blocking funding for research that might otherwise underpin environmental regulations. For example, EPA refused new grants for climate research.[24]

2.4 Research hindrance

The Trump administration also restricted research in other ways. For example, in September 2018, DOI announced plans to destroy records relating to several of its program areas, including "biological resources and marine conservation."[25] The records included, among other things, data on the size and location of various fish and wildlife populations that is used in researching species health.[25] In addition to limiting access to data needed for research, DOI also interfered with research processes. For example, scientists at DOI's U.S. Geological Survey were directed not to model the impacts of climate change beyond 2040, presumably because the worst impacts are expected to occur in the second half of the century.[26]

In total, in the time between President Trump's election and President Biden's inauguration, there were 43 documented examples of research hindrance. The number of incidents rose by 157% from 2017 to 2018—the largest year-on-year increase in any category recorded in the Silencing Science Tracker—before dropping in 2019 and then increasing to 2017 levels again in 2020. Again, many of the actions taken furthered the Trump administration's deregulatory agenda, including by obscuring the harms caused by climate change and thus making it easier to justify the weakening of climate regulations. Other actions appear to have been intended to help President Trump politically. For example, during the COVID-19 pandemic, Trump administration officials pressured the Food and Drug Administration to approve new vaccines and treatments before they had been fully tested.[27] President Trump had previously accused the "deep state, or whoever, over at FDA" of intentionally slowing work to hinder his chances of re-election.[27] His supporters could easily have interpreted this to mean that FDA scientists were pursuing their own (anti-Trump) agenda and thus should not be trusted to deliver impartial advice about COVID-19 or other issues.

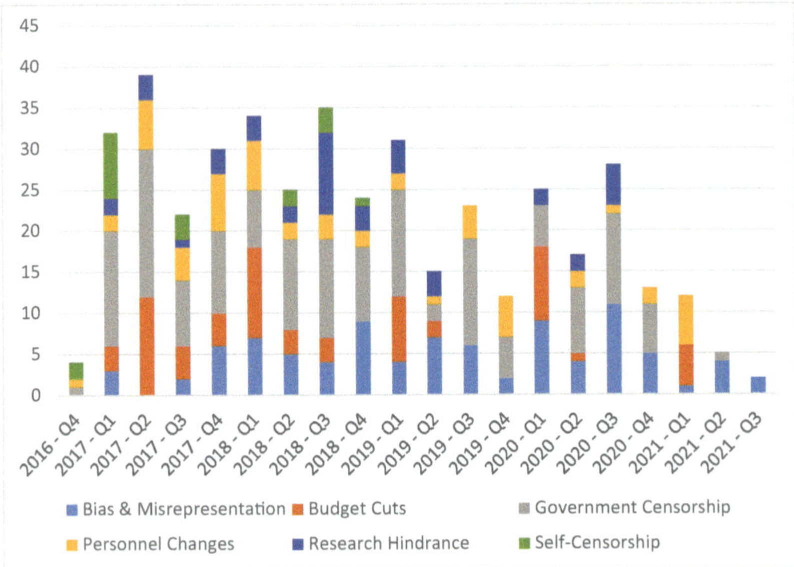

Federal Anti-Science Actions by Quarter (11/08/16 – 01/20/21)

2.5 Bias and misrepresentation

Where the Trump administration could not block the conduct or publication of research (e.g., because it was been mandated by law), administration officials often engaged in bias and misrepresentation, undermining or simply dismissing research findings that did not support its agenda. One notable example was the administration's response to the Fourth National Climate Assessment, which officials falsely claimed was "not data driven" and only modelled "the most extreme scenario," rendering it untrustworthy; President Trump simply declared: "I don't believe it."[28]

President Trump took a similar approach to politically inconvenient COVID-19 research. For example, when research called into question President Trump's claims that hydroxychloroquine was an effective treatment for COVID-19, the former President accused the researchers of intentionally skewing the results by giving the drug to "very sick people··· that were ready to die."[29] President Trump did not point to any evidence to support these claims, but nevertheless labelled the research as "false," and suggested it was politically motivated because that researchers were "obviously not friends of the administration."[29] Again, his supporters were led to believe that scientific research is easily co-opted, and scientific facts open to interpretation.

Federal Anti-Science Actions by Agency (11/08/16 – 01/20/21)

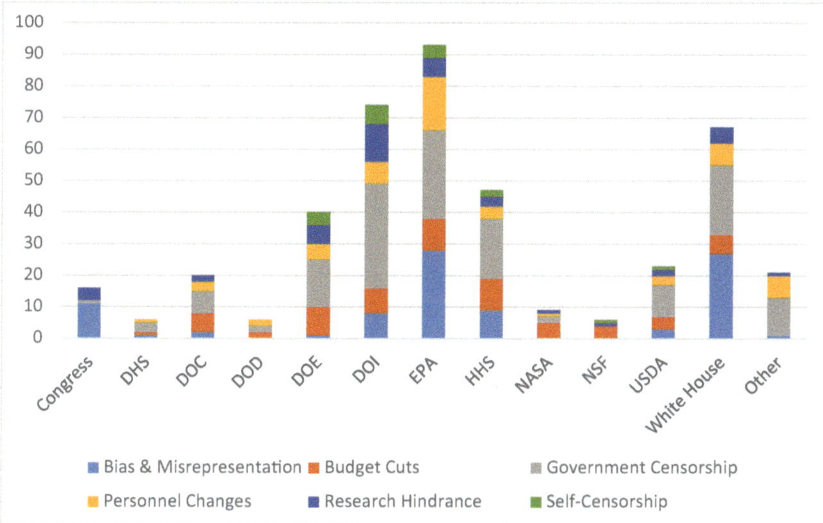

In total, in the period between President Trump's election and when he left office, there were 85 documented instances of bias and misrepresentation involving actors from Congress, the White House, and nine executive agencies. Government actors appear to have felt increasingly emboldened to engage in such behavior during the Trump presidency. Instances of bias and misrepresentation doubled from 2017 to 2018, before stabilizing in 2019 and then increasing again in 2020. The increase may be partly attributable to the Trump administration's widespread scientific censorship, which has limited public access to information that calls officials' views into question. Both government censorship and bias and misrepresentation became increasingly prevalent during the COVID-19 pandemic. In 2020, there were three times as many instances of bias and misrepresentation involving COVID-19 as there were for the next largest category, climate change. (In all other years, climate misinformation was the biggest category of bias and misrepresentation.) Moreover, as a result of other anti-science actions taken by the Trump administration, there were fewer federal scientists to advise and potentially constrain officials during the later years of the administration.

Regardless of the cause, the Trump administration's bias and misrepresentation played neatly into their attempts to dismantle science-based regulations, such as at EPA (where scientists' advice was restricted[30] or outright disregarded[31]) and DOI (which used faulty science to justify deregulation[32]) as well as other agencies like OMB (which instituted guidelines to limit how science can be used by regulatory agencies[33]).

2.6 Agencies affected

These problems were widespread throughout the federal government during the Trump presidency. Anti-science behavior was documented at 23 federal bodies, including, unexpectedly, several agencies not highly focused on scientific research (e.g., the Department of Justice and Federal Communications Commission). Nevertheless, research agencies have borne the brunt of the attacks on science, with the largest number recorded at EPA (93 or 22% of the total) and DOI (74 or 17% of the total).

3. The Biden administration's approach to science

During his campaign and since being elected, President Biden has repeatedly stressed that his administration will "listen to science."[34] He began putting his words into action even before taking office, for example, when he appointed Eric Lander as presidential science advisor.[35] In comparison, Trump did not appoint a science advisor until nearly two years into his term, later than any first-term president since at least 1976.[36] At the same time as he appointed Lander, President Biden also announced that the White House Office of Science and Technology Policy would become a Cabinet-level agency, giving it significantly more influence in administration decision-making.[37]

Building on these early actions, on his first day in office, President Biden issued an executive order declaring that, when combating climate change, "the Federal Government must be guided by the best science and be protected by processes that ensure the integrity of Federal decision-making."[38] One week later, President Biden issued an executive memorandum on "Restoring Trust in Government Through Scientific Integrity and Evidence-Based Policymaking":

> *It is the policy of my Administration to make evidence-based decisions guided by the best available science and data ··· Scientific findings should never be distorted or influenced by political considerations. When scientific or technological information is considered in policy decisions, it should be subjected to well-established scientific processes, including peer review where feasible and appropriate, with appropriate protections for privacy. Improper political interference in the work of Federal scientists or other scientists who support the work of the Federal Government and in the communication of scientific facts undermines the welfare*

of the Nation, contributes to systemic inequities and injustices, and violates the trust that the public places in government to best serve its collective interests.[39]

At President Biden's direction, in May 2021, the White House Office of Science and Technology Policy convened an interagency taskforce to assess the effectiveness of federal agencies' scientific integrity policies and recommend reforms.[40] That is an encouraging first step but significantly more work will be needed to undo the damage done by the Trump administration.

As well as broad efforts to restore the role of science in federal decision-making, the Biden administration will also need to undo many of the individual anti-science actions taken during the Trump presidency. For example, the scientists who left federal employment will need to be replaced, the policies requiring political review of research grants and publications will need to be changed, the federal advisory committees that were disbanded will need to be re-established, and the committees that continue to exist will need to be reviewed to ensure their members are qualified and independent. There is a lot to do and, so far, the Biden administration has only just scratched the surface. At the time of writing—nine months into the Biden presidency—the administration had only undone eleven of the Trump-era anti-science actions recorded in the Silencing Science Tracker. To be fair, some actions cannot be directly undone (e.g., the administration's repeated questioning of climate science). And, in other areas, there have been some encouraging steps forward. For instance, in March 2021, EPA Administrator Michael Regan dismissed all members of the agency's Scientific Advisory Board due to "process irregularities" in appointments during the Trump administration.[41] New board members were appointed in August 2021 after extensive conflict of interest and impartiality tests.[42]

Further work by the Biden administration could be hampered by distrust of science within some segments of the American public. In a survey conducted by the Pew Research Center in April and May 2020, only 39% of respondents said they had a "great deal" of confidence in scientists to act in the best interests of the public.[43] While that was up slightly from 2019 levels, when 35% of survey respondents said they had a "great deal" of confidence in scientists, Pew found growing partisan difference in trust levels.[43] In the 2020 survey, only 27 percent of respondents who identified as Republican expressed a great deal of confidence in scientists, compared to 52% for Democrats.[43] Black and Hispanic Americans have also been shown to have less trust in scientists than White communities.[44]

There can be little doubt that the Trump administration stoked distrust in science in at least some segments of the public. Although previous administrations—both Republican and Democrat—had engaged in anti-science behavior, under President Trump, attacks on science were much more frequent and widespread. They also took on a different flavor. Whereas past presidents consistently upheld the value of scientific research, at least publicly, the Trump administration repeatedly questioned it. Trump administration officials described inconvenient research findings as "untrustworthy" and "unbelievable."[28] Some have even suggested that all research is inherently partisan because science is "a Democrat thing."[23] Those sentiments undermine the perceived value of independent research which could, in turn, encourage greater politicization of science and decrease reliance on it as a basis for environmental and other regulation.

4. Where to from here? Restoring the role of science

Clearly, there is a pressing need for reforms to better protect federal scientists, and restore public trust in the scientific process. Foremost among these is strengthening federal agency scientific integrity policies. Many federal agencies adopted such policies during the Obama administration, with the goal of ensuring that the science they use in decision-making is free from political interference.[45] The policies clearly fell short of their goals during the Trump administration.

The Trump presidency showed that, first and foremost, scientific integrity policies need stronger mechanisms to protect science against political interference. As of this writing, a number of agency scientific integrity policies do not actually prohibit political interference in science.[46] Other policies provide only limited protections, such as at DOI, where only public affairs officers are fully prevented from attempting to engage in political interference.[47] Many policies also need stronger protections for the rights of scientists to communicate their findings, which would allow scientists to freely share scientific information and correct misinformation.[48] And stronger requirements regarding conflicts of interest are needed across the board to help reduce regulatory capture by industry interests and other forms of corruption.[48] Finally, it must be easier and safer for scientists to navigate the scientific integrity complaint process—there is much to be done to clarify the processes and procedures for filing and investigating complaints, implement clear and meaningful penalties for violations, and protect complainants against retaliation.[49]

Several initiatives are currently underway in both the Biden administration and in Congress, working towards these necessary improvements. Unfortunately, on both fronts, progress has been slower than one would like. As mentioned above, following a January 2021 Presidential Memorandum, the Biden administration convened a task force to review and revamp agency scientific integrity policies, but as the writing of this chapter, the task force is already several months behind its initial deadlines. Meanwhile, in Congress, a Scientific Integrity Act was first proposed in 2017 and has been reintroduced several times since then, but does not yet appear to be close to passage.

Despite the slow movement, it is important to remember that the tide is turning in a more pro-science direction. Unfortunately, history has shown that—at some point—the pendulum will likely swing back, at least to some degree. It is imperative that we use this time to institute stronger protections for federal science.

Data availability: All data analyzed in this study are available online at https://climate.law.columbia.edu/Silencing-Science-Tracker.

References

1. Climate Science Legal Defense Fund. https://www.csldf.org/ [https://perma.cc/Z8W7-46BC]. Accessed October 25, 2021.
2. Sabin Center for Climate Change Law. https://climate.law.columbia.edu/ [https://perma.cc/X2L5-D47K]. Accessed October 25, 2021.
3. Envtl. Prot. Agency Office of Inspector General. Further Efforts Needed to Uphold Scientific Integrity Policy at EPA. https://www.epa.gov/sites/production/files/2020-05/documents/_epaoig_20200520-20-p-0173.pdf [https://perma.cc/H8PZ-CA92]. Published May 20, 2020. Accessed October 25, 2021.
4. Science Debate. 2016 Presidential Q&As, https://sciencedebate.org/sciencedebate-presidential-2016.html [https://perma.cc/4DH2-EARH]. Accessed November 1, 2021.
5. The Sabin Center for Climate Change Law, The Climate Science Legal Defense Fund (herein after referred to as "Sabin Center & CSLDF"). Climate Change Information Removed from USDA Website. https://climate.law.columbia.edu/content/climate-change-information-removed-usda-website [https://perma.cc/HA84-P3H6]. Accessed October 25, 2021. Sabin Center & CSLDF. Climate Change Resources Removed from DOE Website. https://climate.law.columbia.edu/content/climate-change-resources-removed-doe-website [https://perma.cc/E26W-JVG6]. Accessed October 25, 2021. Sabin Center & CSLDF. Climate Change Information Removed from DOI Website. https://climate.law.columbia.edu/content/climate-change-information-removed-doi-website [https://perma.cc/4VR7-LJZZ]. Accessed October 25, 2021.
6. Sabin Center & CSLDF. References to "Climate Change" Removed from EPA Regulatory Impact Analysis. https://climate.law.columbia.edu/content/references-climate-change-removed-epa-regulatory-impact-analysis-0 [https://perma.cc/59N3-PPP9]. Accessed October 25, 2021.
7. Sabin Center & CSLDF. Studies Showing Damages from Climate Change Buried by USDA. https://climate.law.columbia.edu/content/studies-showing-damages-climate-change-buried-usda [https://perma.cc/9CYN-H7GP]. Accessed October 25, 2021.
8. Envtl. Prot. Agency, supra note 12, at 17.

9. Sabin Center & CSLDF. NSC Directorate for Global Health Security and Biodefense Dissolved. https://climate.law.columbia.edu/content/nsc-directorate-global-health-security-and-biodefense-dissolved [https://perma.cc/S7V7-RZ2T]. Accessed October 25, 2021.
10. Sabin Center & CSLDF. DOI Scientists Involuntarily Resigned. https://climate.law.columbia.edu/content/doi-scientists-involuntarily-reassigned [https://perma.cc/Y3A2-WYK2]. Accessed October 25, 2021.
11. Sabin Center & CSLDF. Fewer Scientists Employed by USGS During Trump Presidency. https://climate.law.columbia.edu/content/fewer-scientists-employed-usgs-during-trump-presidency [https://perma.cc/8TNU-2CML]. Accessed October 25, 2021.
12. Sabin Center & CSLDF. Fewer Scientists Employed by EPA During Trump Presidency. https://climate.law.columbia.edu/content/fewer-scientists-employed-epa-during-trump-presidency [https://perma.cc/6J25-TV3E]. Accessed October 25, 2021.
13. Sabin Center & CSLDF. One-Third of Advisory Committees to be Terminated. https://climate.law.columbia.edu/content/one-third-advisory-committees-be-terminated [https://perma.cc/QB3J-FWL5]. Accessed October 25, 2021.
14. Sabin Center & CSLDF. DOE to Terminate Three Energy-Related Federal Advisory Committees. https://climate.law.columbia.edu/content/doe-terminate-three-energy-related-federal-advisory-committees [https://perma.cc/N4HK-L28T]. Accessed November 1, 2021.
15. Sabin Center & CSLDF. Trump Administration Axes Federal Advisory Committee Focused on Invasive Species. https://climate.law.columbia.edu/content/trump-administration-axes-federal-advisory-committee-focused-invasive-species [https://perma.cc/9B2B-ABZ7]. Accessed November 1, 2021.
16. Sabin Center & CSLDF. EPA Environmental Laboratory Advisory Board Axed. https://climate.law.columbia.edu/content/epa-environmental-laboratory-advisory-board-axed [https://perma.cc/AD8X-SZXF]. Accessed November 1, 2021.
17. Sabin Center & CSLDF. EPA Science Advisory Board Unofficially Suspended. https://climate.law.columbia.edu/content/epa-science-advisory-board-unofficially-suspended-0 [https://perma.cc/2WUL-V7QD]. Accessed October 25, 2021.
18. Sabin Center & CSLDF, DOL Advisory Boards Suspended. https://climate.law.columbia.edu/content/dol-advisory-boards-suspended-0 [https://perma.cc/ZJ65-L7BR]. Accessed October 25, 2021.
19. Sabin Center & CSLDF, EPA-Funded Scientists Barred From Serving on Advisory Committees. https://climate.law.columbia.edu/content/epa-funded-scientists-barred-serving-advisory-committees [https://perma.cc/ZFZ7-8E94]. Accessed October 25, 2021.
20. Sabin Center & CSLDF. EPA Science Panel Disbanded. https://climate.law.columbia.edu/content/epa-science-panel-disbanded-0 [https://perma.cc/2ASM-XFS6]. Accessed October 25, 2021.
21. Sabin Center & CSLDF. National Climate Assessment Advisory Committee Disbanded. https://climate.law.columbia.edu/content/national-climate-assessment-advisory-committee-disbanded [https://perma.cc/G86Y-WPES]. Accessed October 25, 2021.
22. Hourihan, M. Update: In the Age of Trump, Congress Keeps Boosting Science Funding. https://www.aaas.org/news/update-age-trump-congress-keeps-boosting-science-funding [https://perma.cc/T82E-NS3W]. Published December 17, 2019. Accessed October 27, 2021.
23. Sabin Center & CSLDF. Coal Mining Study Paused by DOI. https://climate.law.columbia.edu/content/coal-mining-study-paused-doi [https://perma.cc/6XCE-QDV8]. Accessed November 1, 2021.

24. Sabin Center & CSLDF. EPA Grants Reviewed by Political Appointee. https://climate.law.columbia.edu/content/epa-grants-reviewed-political-appointee [https://perma.cc/AE2R-ZL9N]. Accessed October 25, 2021.
25. Sabin Center & CSLDF. Scientific Data and Records to be Destroyed by DOI. https://climate.law.columbia.edu/content/scientific-data-and-records-be-destroyed-doi-0 [https://perma.cc/W5P8-AN4N] Accessed October 25, 2021.
26. Sabin Center & CSLDF. USGS Scientists Ordered Not to Model Long-Term Climate Impacts. https://climate.law.columbia.edu/content/usgs-scientists-ordered-not-model-long-term-climate-impacts [https://perma.cc/5KDZ-6BF8] Accessed October 25, 2021.
27. Sabin Center & CSLDF. FDA Commissioner Pressured to Approve Coronavirus Treatment Before Clinical Trial Data Fully Analyzed. https://climate.law.columbia.edu/content/fda-commissioner-pressured-approve-coronavirus-treatment-clinical-trial-data-fully-analyzed [https://perma.cc/G27H-PBDZ]. Accessed October 25, 2021.
28. Sabin Center & CSLDF. Accuracy of National Climate Assessment Questioned by Trump Administration. https://climate.law.columbia.edu/content/accuracy-national-climate-assessment-questioned-trump-administration-0 [https://perma.cc/6XTN-9UR4]. Accessed October 25, 2021.
29. Sabin Center & CSLDF. Accuracy of Scientific Study Questioned by President Trump. https://climate.law.columbia.edu/content/accuracy-scientific-study-questioned-president-trump [https://perma.cc/7PLR-K296]. Accessed October 25, 2021.
30. Sabin Center & CSLDF. Use of Science in EPA Air Pollution Programs Restricted. https://climate.law.columbia.edu/content/use-science-epa-air-pollution-programs-restricted-0 [https://perma.cc/FG24-5RKV] Accessed October 25, 2021.
31. Sabin Center & CSLDF. EPA Scientists' Advice Disregarded. https://climate.law.columbia.edu/content/epa-scientists-advice-disregarded [https://perma.cc/L32E-7KWN]. Accessed October 25, 2021.
32. Sabin Center & CSLDF. FWS Decision to Delist Gray Wolf Based on Faulty Science. https://climate.law.columbia.edu/content/fws-decision-delist-gray-wolf-based-faulty-science [https://perma.cc/YH7Z-8RE5]. Accessed October 25, 2021.
33. Sabin Center & CSLDF. New Guidelines on the Use of Scientific Information Issued by OMB. https://climate.law.columbia.edu/content/new-guidelines-use-scientific-information-issued-omb [https://perma.cc/T5GB-3MQC] Accessed October 25, 2021.
34. Berger E. As it turns out, the Biden administration will listen to scientists. ArsTechnica. https://arstechnica.com/science/2021/01/as-it-turns-out-the-biden-administration-will-listen-to-scientists/ [https://perma.cc/E5EM-WQTE]. Accessed Nov. 1, 2021. *See also* Barrow B, Borenstein, S. Biden says his advisers will lead with 'science and truth.' AP News. https://apnews.com/article/joe-biden-technology-francis-collins-022fc771e262e6f1c7e33ffe80e1d37b [https://perma.cc/GW2D-82FM]. Accessed Nov. 1, 2021.
35. Mervis J, Kaiser J. Biden appoints geneticist Eric Lander as science adviser. *Science.* 2021. https://www.science.org/content/article/biden-appoints-geneticist-eric-lander-science-advisor. [https://perma.cc/HHC5-4SVJ]. Accessed Nov. 1, 2021.
36. Reardon S, Witze A. Trump Taps Meteorologist as White House Science Advisor. *Scientific American.* 2018. https://www.scientificamerican.com/article/trump-taps-meteorologist-as-white-house-science-advisor/. Accessed November 15, 2021.
37. Mervis J, Kaiser J. Biden appoints geneticist Eric Lander as science adviser. *Science.* 2021. https://www.science.org/content/article/biden-appoints-geneticist-eric-lander-science-advisor. Accessed November 1, 2021.

38. Executive Order 14008 of January 27. *Tackling the Climate Crisis at Home and Abroad. 86 Federal Register 7619*; 2021. https://www.federalregister.gov/documents/2021/02/01/2021-02177/tackling-the-climate-crisis-at-home-and-abroad. [https://perma.cc/MEQ7-C7HE]. Accessed November 1, 2021.
39. Memorandum of January 27. *Restoring Trust in Government Through Scientific Integrity and Evidence-Based Policymaking*. 86 Federal Register 8845; 2021. https://www.govinfo.gov/content/pkg/FR-2021-02-10/html/2021-02839.htm. [https://perma.cc/LM3V-PFB5]. Accessed November 1, 2021.
40. The White House. The Biden Administration's Scientific Integrity Task Force Seeks Ideas from the American People. https://www.whitehouse.gov/ostp/news-updates/2021/06/25/the-biden-administrations-scientific-integrity-task-force-seeks-ideas-from-the-american-people/ [https://perma.cc/3QQY-DYWN]. Accessed November 1, 2021.
41. Environmental Protection Agency. Administrator Regan Directs EPA to Reset Critical Science-Focused Federal Advisory Committees. https://www.epa.gov/newsreleases/administrator-regan-directs-epa-reset-critical-science-focused-federal-advisory [https://perma.cc/F7YU-H2RV]. Accessed November 1, 2021.
42. Environmental Protection Agency. EPA Announces Selections of Charter Members to the Science Advisory Board. https://www.epa.gov/newsreleases/epa-announces-selections-charter-members-science-advisory-board [https://perma.cc/H254-4YDL]. Accessed November 1, 2021.
43. Funk C, Kennedy B, Johnson C. Trust in Medical Scientists Has Grown in U.S., but Mainly Among Democrats. https://www.pewresearch.org/science/2020/05/21/trust-in-medical-scientists-has-grown-in-u-s-but-mainly-among-democrats/ [https://perma.cc/TZ9N-69TB]. Accessed October 27, 2021.
44. Funk C, Kennedy B, Tyson A. Black Americans have less confidence in scientists to act in the public interest. https://www.pewresearch.org/fact-tank/2020/08/28/black-americans-have-less-confidence-in-scientists-to-act-in-the-public-interest/ [https://perma.cc/RSW9-5HQU]. Accessed October 27, 2021.
45. Memorandum for the Heads of Executive Departments and Agencies on Scientific Integrity. https://obamawhitehouse.archives.gov/sites/default/files/microsites/ostp/scientific-integrity-memo-12172010.pdf [https://perma.cc/8NJQ-LL3L]. Published December 17, 2010. Accessed November 1, 2021.
46. CDC Guidance on Scientific Integrity. https://www.cdc.gov/os/integrity/docs/CDCSIGuide_042516.pdf [https://perma.cc/F45C-LT8B]. Published April 2016. Accessed November 1, 2021.
47. Department of the Interior. Integrity of Scientific and Scholarly Activities. https://www.doi.gov/sites/doi.gov/files/uploads/305_dm_3_final_revised_si_policy_12-16-14.pdf [https://perma.cc/4P6D-KX8V]. Published December 16, 2014. Accessed November 1, 2021.
48. CSLDF. Guides to Scientific Integrity at Federal Agencies. https://www.csldf.org/resource/guides-to-scientific-integrity-at-federal-agencies/ [https://perma.cc/GHQ8-FFSW]. Accessed October 27, 2021.
49. CSLDF. Model Scientific Integrity Policy for Agencies, Universities, and Other Research Institutions. https://www.csldf.org/resource/model-scientific-integrity-policy/ [https://perma.cc/TZS6-CUU2]. Accessed October 25, 2021.

CHAPTER FIVE

Politicization and COVID-19 vaccine resistance in the U.S.

Toby Bolsen[a,]* and Risa Palm[b]
[a]Department of Political Science, Georgia State University, Atlanta, GA, United States
[b]Urban Studies Institute, Georgia State University, Atlanta, GA, United States
*Corresponding author: e-mail address: tobybolsen@gmail.com

Contents

1. The development of vaccines for COVID-19 82
2. How did the COVID-19 virus itself become politicized? 83
3. Politicization of the COVID-19 vaccine 85
4. Emergence of variants and continued politicization 89
5. What does social science research conclude about promoting vaccine acceptance 90
6. What does politicization mean for practitioners of science? 92
References 94

Abstract

Science is frequently used and distorted to advance political, economic, or cultural agendas. The politicization of science can limit the positive impacts that scientific advances can offer when people reject sound and beneficial scientific advice. Politicization has undoubtedly contributed to hesitancy toward uptake of the COVID-19 vaccine. It is urgent for scientists and clinicians to better understand: (1) the roots of politicization as related to COVID-19 vaccines; (2) the factors that influence people's receptivity to scientific misinformation in politicized contexts; and (3) how to combat the politicization of science to increase the use of life-saving vaccines. This chapter explores these issues in the context of COVID-19 vaccine resistance in the United States. After briefly describing the development of the vaccines, we describe the ways in which the disease itself became politicized because of statements by political leaders and also by media accounts including social media. We then review the politicization of the vaccine at both national and international scales, variability in public acceptance of the vaccines in the United States, and response to the emergence of variants. The next section summarizes social science findings on overcoming vaccine resistance, and the concluding section outlines some of the lessons of the politicization of the disease and the vaccine for health practitioners and life scientists.

Science is frequently used and distorted to advance political, economic, or cultural agendas. In this chapter, we focus on the politicization of science, or its use to advance a particular political agenda. Politicization can occur partly

Progress in Molecular Biology and Translational Science, Volume 188
ISSN 1877-1173
https://doi.org/10.1016/bs.pmbts.2021.10.002

because science "is inherently uncertain... the bottom line is that science is easy to challenge because uncertainty always exists and questioning extant knowledge is part of the research process"[1 (p615)]. One of the primary ways that science is politicized is the emphasis of inherent uncertainties to plant doubt among the public over the existence of any scientific consensus.[2] The politicization of science can prevent scientific advances from occurring and/or limit the positive impacts that ensue when people reject sound and beneficial scientific advice.

Politicization has undoubtedly contributed to hesitancy toward COVID-19 vaccination. It is urgent for scientists and health practitioners to better understand: (1) the roots of politicization as related to COVID-19 vaccines; (2) the factors that influence people's receptivity to scientific misinformation in politicized contexts; and (3) how to combat the politicization of science to increase the use of life-saving vaccines. This chapter explores these issues in the context of COVID-19 vaccine resistance in the United States. After briefly describing the development of the vaccines, we describe the ways in which the disease itself became politicized because of statements by political leaders and media accounts, including social media. We then review the politicization of the vaccine at both national and international scales, variability in public acceptance of the vaccines in the United States, and response to the emergence of variants. The next section summarizes social science findings on overcoming vaccine resistance, and the concluding section outlines some of the lessons of the politicization of the disease and the vaccine for health practitioners and life scientists.

1. The development of vaccines for COVID-19

The WHO declared COVID-19 a global pandemic on March 11, 2020, leading the United States to declare COVID-19 a national emergency 2 days later. Almost immediately, on March 16, phase I of clinical trials began for "an mRNA vaccine targeting spike protein" designed by the National Institutes of Health and Moderna (m-RNa-1273); Pfizer and BioNTech announced the start of a phase I/II trial of their mRNA vaccine (BNT162) on May 5, 2020. The US Food and Drug Administration (FDA) issued emergency use authorization of the Pfizer-BioNTech vaccine on December 11, 2020, and the Moderna vaccine on December 18, 2020. Emergency use authorization was granted to the Johnson & Johnson vaccine on February 27, 2021. The Pfizer-BioNTech vaccine gained full FDA approval on August 23, 2021.

The rapid development and testing of these mRNA vaccines was made possible by previous decades of work and many millions of dollars of investment in immunology and vaccine research programs by both the federal government, through the National Institutes of Health, and private foundations such as the Bill and Melinda Gates Foundation.[3] Previous work on the severe acute respiratory syndrome (SARS), Middle East respiratory syndrome (MERS), influenza, and HIV viruses laid the foundation for strategies to neutralize these viruses, including development of antibody cocktails. The technology underlying the Pfizer/BioNTech and Moderna/NIH vaccines emerged over the last decade from university scientists working on HIV and influenza vaccines, which then sparked Zika, Ebola, and coronavirus vaccine programs at the NIH and in the pharmaceutical industry. The Moderna mRNA vaccine was made as a collaboration with the NIH's Vaccine Research Center, funded by U.S. taxpayers since 1997 to create vaccines against deadly viruses and other human diseases.[4,3]

2. How did the COVID-19 virus itself become politicized?

The politicization of health-related issues occurs when political identities or cues become integrated into the public discourse surrounding any topic.[5,6] This can involve news outlets emphasizing political controversy over an issue that then colors partisans' evaluations of novel scientific information. The roots of politicization surrounding COVID-19 vaccines can be attributed, at least in part, to (1) statements and visual cues by leading politicians, (2) media coverage about the nature of the virus; and (3) partisan divisions in public discourse about vaccination.[7,8]

(1) *Statements and visual cues by leading politicians and the CDC*

President Donald Trump initially downplayed the threat posed by COVID-19 and compared it to the flu in public remarks.[9] The President also labeled it a "new hoax" [and said that] "the Democrats are politicizing the coronavirus—they're politicizing it."[10] President Trump also promoted claims that the virus could be combatted by injecting or drinking disinfectant or bleach, and promoted hydroxychloroquine as a cure.[11,12] He also appeared in public without a mask and criticized his rival for the presidency, Joe Biden, for wearing one: "Did you ever see a man that likes a mask as much as him?... If I were a psychiatrist, I'd say this guy has some big issues."[13]

Politicization thus was "baked into the context of the emergent coronavirus... From the earliest alarm, Republican politicians followed Trump's lead in publicly downplaying the threat, while Democrats responded with more concern, exhibiting different public cues."[9(p969),14] Further, portions of the American public were influenced by President Trump's messages on Twitter suggesting that the virus was a hoax, was not serious, or that unproven therapies should be used.[15–17]

In the United States, the primary agency "recognized as the nation's premiere health promotion, prevention, and preparedness agency" is the Centers for Disease Control and Prevention or the CDC.[18] Established in 1946, this agency has made significant contributions to public health in the nation through its studies of communicable and non-communicable diseases. Unfortunately, in the context of the COVID-19 pandemic, the FDA conveyed mixed messages to the public about the risks of the disease and modes of prevention. As Michael Lewis described it, the agency's relationship to disease control "had changed in ways that eliminated its need for bravery. It had begun a descent. It had replaced the flowers on its porch with fake ones and hoped no one would notice."[19(p290)] Guharoy and Krenzelok suggest that the CDC's "unquestionable record was tarnished by technical blunders, lack of leadership, and contradictory messages throughout the pandemic."[20 (p4)] They cite as examples the failure to provide a COVID-19 test kit during early stages of the pandemic and its probable acquiescence to pressure from the Trump administration "to encourage the use of unvalidated treatments."

A staff report from the Select Subcommittee on the Coronavirus Crisis commissioned by the US House of Representatives identified 47 instances of government interference: "repeatedly overruling and sidelining top scientists and undermining Americans' health to advance the President's partisan agenda."[21(p1)] Among the instances of such partisan government interference were delays in a CDC travel alert, the blocking of plans to send reusable masks to all US households in April of 2020, the lifting of shutdown orders recommended by the CDC, and the delay and censorship of scientific reports on the nature of the virus.

(2) *Media coverage*

Conflicting partisan messaging has been propagated by various news sources since the beginning of the outbreak. For instance, several popular right-leaning media outlets suggested that the virus was not as severe a health threat as was being portrayed. Instead, they claimed that coverage of the seriousness of the virus was misleading, a conspiracy by the Chinese government to harm the US economy, or a plot by the "deep state" to spread panic and hurt President Trump's chances for re-election. During the COVID-19 outbreak, Fox News was far more likely than CNN or MSNBC to include phrases raising skepticism about the impacts of the virus, with language such as "normal flu," "political weapon," and "flu panic" more prevalent in their coverage from February 1 through April 30, 2020.[7,8,22]

Social media messages compounded politicization of the virus.[23,24] Stories that circulated widely on social media included false claims (such as transmission of coronavirus through mosquito bites), conspiracy theories (the virus is spread by 5G towers) and pseudoscientific health therapies (eating garlic or drinking bleach can cure the disease).[25] The speed and extent of this spread of false information via social media has been characterized as a kind of epidemic itself: a "misinfodemic."[26]

(3) *Partisan differences in response*

In the absence of a vaccine, governments implemented a variety of nonpharmaceutical interventions (NPIs) "such as social distancing, school closures, remote working, restrictions concerning public gatherings, quarantines, hand-washing and the use of masks to slow transmission of the disease."[27(p684)] The success of these NPIs, as with vaccinations, was affected by politicization in media coverage and communication through social media resulting in, among other things, partisan differences in the perceived seriousness of the virus and willingness to comply with NPIs.[28–33]

3. Politicization of the COVID-19 vaccine

Political conflict and coverage surrounding the nature and risk of the virus set the stage for segments of the public to be relatively more receptive to misinformation about COVID-19 vaccines.[34] Attitudes toward receiving a vaccine can be "understood as a continuum ranging from outright refusal to active demand for immediate uptake."[35(p1), 36,37] Since the beginning stages of testing for COVID-19 vaccines, polls indicated that a sizeable percentage of people in the United States were *vaccine resistant*, defined as unwilling to get vaccinated once one became available.[38] Many people were also *vaccine hesitant*, defined as a preference to wait for others have been vaccinated before deciding for themselves.[39,40] It is important to note that vaccine resistance/hesitancy has not been static but instead has been influenced by information, the current state of the epidemic, and a variety of individual-level characteristics.[41]

(1) *Coverage of vaccine development and approval*

On May 15, 2020, President Trump announced Operation Warp Speed, a public-private partnership with $10 billion in funding from Congress to enable faster development and approval of a vaccine for COVID-19.

"That means big, and it means fast," Trump stated, "A massive scientific, industrial and logistical endeavor unlike anything our country has seen since the Manhattan Project."[42] The main goal, according to the Department of Health and Human Services, was to "produce and deliver 300 million doses of safe and effective vaccines with the initial doses available by January 2021."[43] This provided support for pharmaceutical companies to conduct research and development on vaccine candidates, aid for manufacturing and storage of pre-approved vaccines, and coordination across other federal agencies to accelerate the process. Yet misinformation about COVID-19 vaccines was already widespread and "a more fraught scenario for science communication is hard to imagine: a novel vaccine, probably fast-tracked, in the middle of a highly politicized and badly mishandled pandemic."[44] The Trump administration's title for this initiative likely played into the hands of anti-vaccine groups by allowing them to frame COVID-19 vaccine trials as prioritizing speed over safety. Anthony Fauci, Director of the National Institute of Allergy and Infectious Diseases stated, "I'm a little concerned by that name because it can imply… that you're going so fast that you're skipping over important steps and are not paying enough attention to safety, which is absolutely not the case."[42]

Operation Warp Speed coincided with a US presidential campaign and election, during which the safety and efficacy of any emergency-authorized COVID-19 vaccines was the subject of media coverage driven by a focus on conflict and the partisan implications of vaccinations. China and Russia had been sharply criticized for authorizing emergency use of COVID-19 vaccines outside of clinical trials, which led to "worry that political forces—the U.S. presidential election on 3 November, nationalistic pride to 'win' a race, the need to resuscitate economies—could lead to premature and dangerous approvals…by the U.S. Food and Drug Administration."[45] President Trump also had stated that emergency authorization for a COVID-19 vaccine prior to Election Day would likely help his chances for re-election.[46] The *New York Times* published an article voicing concerns about an October Surprise in which the FDA might rush a vaccine's approval to please the President.[47] In addition, the vaccine was the subject of political debates over the benefits of government-mandated vaccinations, concerns over "immunization cards" that people might be required to carry, and a loss of individual rights by allowing government to make personal health decisions.[48,49] This politicization of the vaccine's approval process, and the idea that the vaccine might be released before the election, reduced the public's confidence in its safety and their intentions to get vaccinated.[50]

Vaccine development also was politicized on a national scale as China, Russia and the United States raced to be the first with an effective vaccine. Government agencies and leading news sources attributed this to "vaccine nationalism." For example, comparing coverage of vaccine development in the Global Times (China) and the New York Times (US), Abbas found that Chinese state media started "a misinformation campaign against American vaccines" while US news sources raised doubts about the strength and effectiveness of the Chinese Sinovac vaccine.[51]

(2) *Variability in vaccine acceptance/resistance*

Politicization surrounding COVID-19 vaccines contributed to hesitancy, resistance, and opinion polarization. In June 2020, polls showed that about 34% of the U.S. public would accept a COVID-19 vaccine, and large differences existed in opinions between Republicans Democrats.[52,53,41; 1] Those who intended to vote for President Trump were 35% more likely to say that they would not get vaccinated for COVID-19.[54] As vaccine access expanded in the spring of 2021, the proportion of Americans stating that they intended to get vaccinated rose, but still lagged among Republicans.[55] As Allcott et al. summarized it: "partisanship is a primary driver of attitudes about the pandemic and self-reported behaviors."[28(p4)]

Republicans were more likely than Democrats to believe anti-vaccine misinformation.[34] This reflected a growing skepticism among conservatives toward the scientific community in general, in part due to the rise of right-wing populist messages that pit "ordinary people" against "corrupt elites."[56,57] This distrust of experts was associated with the rejection of scientific messages about vaccines.[58] Populist rhetoric may have primed anti-intellectual considerations that made public health messages less impactful among conservatives.[59]

In this case, attention to the statements and endorsements made by trusted elite sources led to high levels of polarization. For instance, a study was conducted in late March 2021 on the effect of exposure to one of two short videos and statements endorsing COVID-19 vaccines—either from President Joe Biden or Former President Donald Trump.[55] Republicans hearing a message urging vaccination from Trump were more likely to state

---

[1] Several other demographic characteristics have also been associated with reported willingness to accept the COVID-19 vaccine. In general, people with less education, income, and who live in rural areas were less likely to get vaccinated for COVID-19; and, at least in the U.S., women and minorities were significantly more likely to say they would not get vaccinated.[40] Americans characterized as having high levels of religiosity are also both more likely to be misinformed about COVID-19 and also to refuse a vaccine.[111]

they would get vaccinated compared to a control group, but were less likely to state they would get vaccinated if the message came from President Biden, the out-party leader. Even for those seeking accurate information, differences in the *perceived credibility* of partisan sources could have driven the observed belief polarization occurring after exposure to identical messages from opposing partisan elites.[60] The key take-away points are that people's underlying goals, or motivations, vary across contexts and unvaccinated Republicans would be more likely to accept a COVID-19 vaccine after exposure to a pro-vaccination message endorsed by a Republican leader. Studies have also shown that people conform to the opinions of others living in the same area: a study showed that people living in areas with higher degrees of polarization were more likely to conform with others of their in-group with respect to health-behavior responses to the COVID-19 crisis.[61,2]

Human decision-making and information processing is influenced by *motivations* that color how people evaluate scientific arguments, evidence, and political information.[62–64] COVID-19 was "a disease that quickly became politicized in the United States, [and] individuals are vulnerable [to vaccine misinformation] because they tend to rely on identity affirmation rather than a systematic assessment of information" in these settings.[55] A reliance on "identity affirmation" in processing scientific information means that people often are motivated to evaluate new information in a way that protects their prior beliefs, worldviews, and social identities: when forming beliefs and making decisions, people may seek out information that supports their existing views (confirmation bias), selectively avoid or counter argue against information that challenges and existing belief or identity (disconfirmation bias), and evaluate information that is congruent with an existing belief, worldview or identity as stronger than opposing arguments (prior attitude effect).[65,66]

Politicization of the vaccine also interacted with the spread of conspiracy theories about the vaccine.[3] A conspiracy theory is an unconfirmed explanation for an event that involves a small group of actors engaging in a harmful deception for personal benefit at the expense of the collective.[67] In the case of COVID-19, conspiracy theories singled out "China, Russia, Bill

---

[2] It should be noted that this effect is conditional on the severity of the pandemic and highlights that there are ways to vitiate "biases" in information processing and reasoning.

[3] A sizeable literature has shown that a general conspiratorial predisposition exists, independent of partisanship, that causes people to be receptive to conspiracy theories and claims and more skeptical of scientific counterclaims and recommendation.

Gates, Democrats, the 'deep state', and the pharmaceutical industry, to name a few."[68(p2)] A national survey of US adults in June 2020 showed that conspiracy theories endorsed by partisan figures have a stronger impact on audiences' levels of misinformation about COVID-19 compared to medical misinformation about the transmissibility of the virus.[68] Individuals who are more open to believing in conspiracy theories, especially on issues where there is politicization and partisan polarization, tend to be more resistant to scientific information, more open to science politicization, and less willing to engage in pro-social health behaviors.[69–72]

4. Emergence of variants and continued politicization

The SARS-CoV-2 virus, like all other viruses, change and evolve over time. It remains uncertain how the virus might change with respect to its infectiousness, virulence, and immune evasion.[73] At this point, the World Health Organization (WHO) has four "variants of concern" identified in the United Kingdom (alpha, September 2020), South Africa (beta, May 2020), Brazil (gamma, November 2020) and India (Delta, October 2020).[74] The WHO has also designated five other variants as "variants of interest" in addition to numerous other variants that are designated "Alerts for Further Monitoring." The list is amended frequently as new mutations are identified and documented.

As of August 2021, the delta variant had become the most prevalent because of a mutation that made it more infectious than the original virus. Because of its increased ability to spread among human populations, "almost all current viruses are descended from it ... evidence is accumulating that infected people not only spread the virus more efficiently, but also faster, allowing the variant to spread even more rapidly."[73(p846)] With large numbers of people unvaccinated, the variants of the virus could become even more deadly immune to current vaccines.[75,76]

Despite a barrage of media reports on renewed spikes in deaths and hospitalizations due to COVID-19, and warnings from various celebrities and political leaders, the response to the adoption of public health measures continues to be polarized. In summer 2021, 52% of Republicans said that they had received at least one dose of the vaccine or planned to get vaccinated as soon as possible as opposed to 88% of Democrats.[77] Further evidence of this division is the fact that 17 of the 20 States with the lowest vaccination rates in summer 2021 were carried by Donald Trump in the 2020 Presidential

Election.[78] Politicization surrounding mask mandates in public spaces including classrooms and mandatory vaccinations or testing for some exemplify these ongoing divisions.

Reluctance to abide by mandatory masking edicts in public venues and classrooms became politicized early in the course of the pandemic as advice from the Centers for Disease Control and public health experts originally discouraged masks for people not showing symptoms.[79] Concerned about a possible shortage of masks, experts including Dr. Anthony Fauci, director of the National Institute of Allergy and Infectious Diseases and Dr. Jerome Adams, former surgeon general argued that face masks should be reserved for front-line medical workers.[80] When the message changed and some school districts and government authorities began mask mandates, protests erupted arguing that mask requirements impinged on personal freedom.[81] In states such as Florida, Texas and Arizona, state leaders banned mask mandates for schools, while the University System of Georgia regents warned of disciplinary action for faculty members requiring COVID mitigation measures such as distance learning or mandatory masks in classrooms.[82,83] In each such case, the anti-mask position was backed by a Republican governor and/or Republican-dominated legislature arguing that such mandates impinge on parental rights to make decisions about the health care of their children.[84]

The announcement by President Biden on September 10, 2021, that employers with 100 workers or more require their employees to get vaccinations or weekly COVID testing, was met with sharp criticism from Republicans who saw this move as "unconstitutional," "un-American" and "an assault on freedom and liberty."[85–87] At the same time, the chair of the Democratic National Committee, Jaime Harrison, enthusiastically supported this mandate, stating that the Democrats are "moving forward with protecting the American people" while the Republican response is "crazy."[88]

5. What does social science research conclude about promoting vaccine acceptance

To control the spread of COVID-19 through vaccination, it is imperative that communication strategies to overcome hesitancy and resistance among unvaccinated individuals and groups be developed. One strategy is to convey to the public the dangers of viral infection of themselves and others, and the value of vaccination to prevent disease and death, actions that might lead to decreased vaccination rejection. To better understand the causal impact of messages about COVID-19 vaccines, we conducted a controlled survey experiment in May 2020 that randomly assigned 1123 respondents to one of six different message conditions.[4] The messages (treatments) highlighted positive or negative considerations about the safety of the coronavirus vaccine, information about the likelihood that other Americans would get vaccinated when a vaccine became available, and politicization of vaccination as evidenced by suggestions that liberals sought to control individuals' personal lives or by President Trump's unsupported statements that vaccine approval was rushed for political gain by Democrats and other liberals. We found that those given a message emphasizing the safety and effectiveness of the vaccine reported increased intentions to get vaccinated compared to the control group of respondents who did not read or receive any information (treatment) but were simply asked about their intentions to get vaccinated. We also found that messages that most other Americans were *not* willing to get vaccinated decreased reported intentions to get vaccinated, demonstrating the powerful impact that communicating descriptive social norms can have on individuals. In addition, messages that suggested the vaccine was being rushed for approval prior to the Presidential election decreased reported intentions to get vaccinated.[35]

Another study conducted in December 2020 and January 2021 on 24,682 unvaccinated Americans involved asking respondents to read one of five randomly assigned messages. These included whether the vaccine was recommended by the respondent's personal physician or "most scientists," two moral frames that appealed either to patriotism ("doing what is right for one's country") or preventing harm to themselves and others, or to a descriptive social norm treatment that communicated whether or not most people in one's social network were likely to take the vaccine.[89] The results showed that exposure to every message condition, except for patriotism, significantly reduced vaccine resistance. Further, the large sample offered an opportunity to explore how different subgroups responded to the different messages. Few differences in the effects of these messages on different partisan subgroups were found. These studies indicated that "multiple messaging strategies can reduce vaccine resistance." However, a key question that remains is "when are these communications most effective, and for whom?"

[4] This was a convenience sample recruited and remunerated through Amazon's Mechanical Turk (MTurk) platform.

Scientific messages that communicate the degree of scientific consensus on any issue can also have a powerful impact on audiences and may serve as one way to combat vaccine resistance.[90,91] One study that looked at perceptions of the safety of childhood vaccines found that agreement with the statement that "90% of medical scientists agree that vaccines are safe" was the strongest predictor of support for vaccines.[92] Shifting individuals' perceptions about the level of consensus among expert scientists about the safety and efficacy of the coronavirus vaccine might ultimately lead to a cascade of pro-social health behavior, such as getting vaccinated.

Other research has explored the extent to which exposing individuals to information that warns against and refutes, or "pre-bunks," vaccine misinformation prior to actual exposure to the misinformation reduces or eliminates the effect it can exert on audiences in politicized contexts.[93] This builds on efforts to combat other forms of scientific misinformation by making respondents aware of such claims but then immediately refuting them in order to establish an initial belief that the misinformation is not to be trusted.[94–97] Although pre-bunking is a possible way of combatting vaccine misinformation prior to its politicized embedding in the public consciousness, it is not always possible to preempt the rise and spread of misinformation, especially in a transformed digital media environment.[98–100] In such cases, providing individuals with factually accurate and corrective information often can reduce political and scientific misperceptions.[101,102] In a study performed in the United States and United Kingdom, Carey et al. demonstrated that corrections reduced targeted misperceptions on incorrect beliefs—that hydroxychloroquine and antibiotics can cure COVID-19, that the virus was a bioweapon created by China or created by Bill Gates—but did not have a persistent and lasting impact on the prevalence of those same misperceptions even when survey respondents were repeatedly exposed to corrective information. On the positive side, providing factually corrective information increased the accuracy of respondents' perceptions about COVID-19 and reduced the prevalence of these misperceptions in the short-term; however, the effects of this information were "disappointingly ephemeral" on those respondents and completely disappeared over time.[103]

6. What does politicization mean for practitioners of science?

Online social media has facilitated the spread of rumors, conspiracy theories and misinformation at a speed and intensity unprecedented in history.[104] When combined with the politicization of science, the results can be disastrous for objective reasoning in individual and collective decision-making.[105] What can science practitioners do to combat the deleterious impacts of politicization in the specific case of resistance to the COVID-19 vaccine? In addition to individual activities on the part of scientists themselves, there are several take-aways for life scientists and health professionals.

First, we need to recognize that science is politicized, but that such politicization has been the case throughout much of human history. From the conviction of Socrates (470–339 BC) for corrupting youth with his teachings, to Roger Bacon (1214–94) being imprisoned for finding logical inconsistencies in the Bible, to Galileo Galilei's arrest for his refusal to accept the Church's position on the nature of the solar system, scientists have struggled for independence and intellectual freedom. An account written in 2009 reports "threats from all sides of the political spectrum over the years" including restricting funding, dictating the curriculum in science classes, controlling the composition of scientific advisory committees, and reviewing or censoring scientific publications."[106(p4)] Although the speed and efficiency of transmission of science skepticism has changed with the democratization of information, the tendency to resist ideas and inventions that would be beneficial to the world has a long history.[107,108] Understanding the context of skepticism about science should increase patience with short-term setbacks and also the need for persistence even under challenging situations.

Second, scientists need to be aware of how their communications affect people and tailor their communication styles to maximize effectiveness. Decades of research on effective communication of science suggests a few principles (1) effective communication leading to constructive action requires that the communicator has earned the trust of the audience, not through status or titles, but instead through mutual respect; (2) scientific information is more likely to be acted on if technical details are accompanied by content to which the audience relates, including storytelling; (3) information should be clear, consistent, understandable and actionable.[109] Just as social media has been used in the COVID pandemic to communicate misinformation, so too can it be a tool used by the science community to correct that misinformation and to enhance trust in scientific and medical research findings.

Finally, there is a need for more collaborative cross-disciplinary work to promote rapid adoption of important scientific findings. The complexity of the COVID-19 pandemic and its politicization at local, national, and international scales points to the importance of multifaceted approaches to understanding its disease-related, pharmaceutical, economic, geographic, cognitive, and psychological components simultaneously.[110] The improvement of

COVID outcomes will only take place when global expertise gleaned from all relevant disciplines are marshaled. Life scientists and healthcare professionals will play a key role, but their work will be enhanced if they work in large-scale teams with social and psychological scientists as well as experts in policy and communication.

References

1. Druckman JN. The crisis of politicization within and beyond science. *Nat Hum Behav.* 2017;1(9):615–617. https://doi.org/10.1038/s41562-017-0183-5.
2. Oreskes N, Conway EM. *Merchants of Doubt: How a Handful of Scientists Obscured the Truth on Issues from Tobacco Smoke to Global Warming.* Bloomsbury Publishing; 2011.
3. Moore JP, Wilson IA. Decades of based research paved the way for today's 'war speed; Covid-29 vaccines. *Stat.* 2021; [https://www.statnews.com/2021/01/05/basic-research-paved-way-for-warp-speed-covid-19-vaccines/. Accessed [September 30, 2021].
4. Dolgin E. The tangled history of mRNA vaccines. *Nature.* 2021;597(7876):318–324. https://doi.org/10.1038/d41586-021-02483-w.
5. Fowler EF, Gollust SE. The content and effect of politicized health controversies. *Ann Am Acad Pol Soc Sci.* 2015;658(1):155–171. https://doi.org/10.1177/0002716214555505.
6. Lupia A. Communicating science in politicized environments. *Proc Natl Acad Sci.* 2013;110(Supplement_3):14048–14054. https://doi.org/10.1073/pnas.1212726110.
7. Hart PS, Chinn S, Soroka S. Politicization and polarization in COVID-19 news coverage. *Sci Commun.* 2020;42(5):679–697. https://doi.org/10.1177/1075547020950735.
8. Motta M, Stecula D, Farhart C. How right-leaning media coverage of COVID-19 facilitated the spread of misinformation in the early stages of the pandemic in the U.S. *Can J Pol Sci.* 2020;53(2):335–342. https://doi.org/10.1017/S0008423920000396.
9. Gollust SE, Nagler RH, Fowler EF. The emergence of COVID-19 in the US: A public health and political communication crisis. *J Health Politics Policy Law.* 2020;45(6):967–981. https://doi.org/10.1215/03616878-8641506.
10. Cook N, Choi M. Trump rallies his base to treat coronavirus as a 'hoax'. *Politico.* 2020; [February 28, 2020. www.politico.com/news/2020/02/28/trump-south-carolina-rally-coronavirus-118269. Accessed [September 30, 2021].
11. Kahane LH. Politicizing the mask: Political, economic and demographic factors affecting mask wearing behavior in the USA. *Eastern Econ J.* 2021;47(2):163–183. https://doi.org/10.1057/s41302-020-00186-0.
12. Mackey TK, Purushothaman V, Haupt M, Nali MC, Li J. Application of unsupervised machine learning to identify and characterise hydroxychloroquine misinformation on twitter. *The Lancet Digital Health.* 2021;3(2):e72–e75. https://doi.org/10.1016/S2589-7500(20)30318-6.
13. LeBlanc P. Trump mocks Biden for wearing mask: 'Did you ever see a man that likes a mask as much as him?'. *CNN.* 2020; [https://edition.cnn.com/2020/09/03/politics/trump-biden-coronavirus- mask/index.html. Accessed [September 30, 2021].
14. Hao F, Shao W, Huang W. Understanding the influence of contextual factors and individual social capital on American public mask wearing in response to COVID–19. *Health Place.* 2021;68:102537. https://doi.org/10.1016/j.healthplace.2021.102537.
15. Hornsey MJ, Finlayson M, Chatwood G, Begeny CT. Donald Trump and vaccination: The effect of political identity, conspiracist ideation and presidential tweets on vaccine hesitancy. *J Exp Soc Psychol.* 2020;88:103947. https://doi.org/10.1016/j.jesp.2019.103947.

16. Ugarte DA, Cumberland WG, Flores L, Young SD. Public attitudes about COVID-19 in response to president Trump's social media posts. *JAMA Netw Open*. 2021;4(2): e210101. https://doi.org/10.1001/jamanetworkopen.2021.0101.
17. Niburski K, Niburski O. Impact of Trump's promotion of unproven COVID-19 treatments on social media and subsequent internet trends: Observational study. *J Med Internet Res*. 2020;22(11):e20044. https://doi.org/10.2196/20044.
18. Centers for Disease Control and Prevention. *Our History*. CDC Healthy Schools; 2021 [https://www.cdc.gov/healthyschools/bam/basics/our-history.htm. Accessed [September 30, 2021].
19. Lewis M. *The Premonition: A Pandemic Story*. UK: Penguin; 2021.
20. Guharoy R, Krenzelok EP. Lessons from the mismanagement of the COVID-19 pandemic: A blueprint to reform CDC. *Am J Health Syst Pharm*. 2021;78 (18):1739–1741. https://doi.org/10.1093/ajhp/zxab209.
21. United States House of Representatives, Select Subcommittee on the Coronavirus Crisis. *The Trump administration's Pattern of Political Interference in the nation's Coronavirus Response*; 2020. https://coronavirus.house.gov/sites/democrats.coronavirus.house.gov/files/10.2.20%20Political%20Interference%20Report%20%281%29.pdf. Accessed [September 30, 2021].
22. Ash E, Galletta S, Hangartner D, Margalit Y, Pinna M. The effect of fox news on health behavior during COVID-19. *SocArXiv*. 2020. https://doi.org/10.31235/osf.io/abqe5.
23. Megget K. Even covid-19 can't kill the anti-vaccination movement. *BMJ*. 2020; m2184. https://doi.org/10.1136/bmj.m2184. Published online June 4.
24. Pian W, Chi J, Ma F. The causes, impacts and countermeasures of COVID-19 "Infodemic": A systematic review using narrative synthesis. *Inf Process Manag*. 2021;58(6):102713. https://doi.org/10.1016/j.ipm.2021.102713.
25. Naeem SB, Bhatti R, Khan A. An exploration of how fake news is taking over social media and putting public health at risk. *Health Info Libr J*. 2021;38(2):143–149. https://doi.org/10.1111/hir.12320.
26. Chou W-YS, Gaysynsky A, Vanderpool RC. The COVID-19 misinfodemic: Moving beyond fact-checking. *Health Educ Behav*. 2021;48(1):9–13. https://doi.org/10.1177/1090198120980675.
27. Soveri A, Karlsson LC, Antfolk J, Lindfelt M, Lewandowsky S. Unwillingness to engage in behaviors that protect against COVID-19: The role of conspiracy beliefs, trust, and endorsement of complementary and alternative medicine. *BMC Public Health*. 2021;21(1):684. https://doi.org/10.1186/s12889-021-10643-w.
28. Allcott H, Boxell L, Conway J, Gentzkow M, Thaler M, Yang D. Polarization and public health: Partisan differences in social distancing during the coronavirus pandemic. *J Public Econ*. 2020;191:104254. https://doi.org/10.1016/j.jpubeco.2020.104254.
29. Gadarian SK, Goodman SW, Pepinsky TB. Partisanship, health behavior, and policy attitudes in the early stages of the COVID-19 pandemic. *PLoS ONE*. 2021;16(4): e0249596. https://doi.org/10.1371/journal.pone.0249596.
30. Simonov A, Sacher S, Dubé J-P, Biswas S. The persuasive effect of fox news: Non-compliance with social distancing during the Covid-19 pandemic. *National Bureau of Economic Research*. 2020;w27237. https://doi.org/10.3386/w27237.
31. Barrios J, Hochberg Y. Risk perception through the lens of politics in the time of the COVID-19 pandemic. *National Bureau of Economic Research*. 2020. https://doi.org/10.3386/w27008.
32. Grossman G, Kim S, Rexer JM, Thirumurthy H. Political partisanship influences behavioral responses to governors' recommendations for COVID-19 prevention in the United States. *Proc Natl Acad Sci U S A*. 2020;117(39):24144–24153. https://doi.org/10.1073/pnas.2007835117.

33. Makridis C, Rothwell JT. The real cost of political polarization: Evidence from the COVID-19 pandemic. *SSRN Journal.* 2020. https://doi.org/10.2139/ssrn.3638373. Published online.
34. Motta M. Republicans, not democrats, are more likely to endorse anti-vaccine misinformation. *Am Politics Res.* 2021;49(5):428–438. https://doi.org/10.1177/1532673X211022639.
35. Palm R, Bolsen T, Kingsland JT. The effect of frames on COVID-19 vaccine resistance. *Front Polit Sci.* 2021;3:661257. https://doi.org/10.3389/fpos.2021.661257.
36. Dubé E, Laberge C, Guay M, Bramadat P, Roy R, Bettinger JA. Vaccine hesitancy: An overview. *Hum Vaccin Immunother.* 2013;9(8):1763–1773. https://doi.org/10.4161/hv.24657.
37. Larson HJ, Jarrett C, Eckersberger E, Smith DMD, Paterson P. Understanding vaccine hesitancy around vaccines and vaccination from a global perspective: A systematic review of published literature, 2007–2012. *Vaccine.* 2014;32(19):2150–2159. https://doi.org/10.1016/j.vaccine.2014.01.081.
38. Lazer D, Ognyanova K, Green J, et al. The COVID states project #47: Update on COVID-19 vaccine attitudes among healthcare workers. *Open Science Framework.* 2021. https://doi.org/10.31219/osf.io/a352z.
39. Simonson MD, Baum M, Lazer D, et al. The COVID states project #45: Vaccine hesitancy and resistance among parents. *Open Science Framework.* 2021. https://doi.org/10.31219/osf.io/e95bc.
40. Lin C, Tu P, Beitsch LM. Confidence and receptivity for COVID-19 vaccines: A rapid systematic review. *Vaccine.* 2020;9(1):16. https://doi.org/10.3390/vaccines9010016.
41. Loomba S, de Figueiredo A, Piatek SJ, de Graaf K, Larson HJ. Measuring the impact of COVID-19 vaccine misinformation on vaccination intent in the UK and USA. *Nat Hum Behav.* 2021;5(3):337–348. https://doi.org/10.1038/s41562-021-01056-1.
42. Weixel N. Some worry 'operation warp speed' plays into anti-vaccination movement's hands. *The Hill.* 2020; [https://thehill.com/policy/healthcare/500213-some-worry-operation-warp-speed-plays-into-anti-vaccination-movements-hands. Accessed [September 30, 2021].
43. US Department of Health and Human Services. *Fact Sheet: Explaining Operation Warp Speed*; 2020 [November 10, 2020. Accessed [September 30, 2021].
44. Zhang S. We don't even have a COVID-19 vaccine, and yet the conspiracies are here. *The Atlantic.* 2021; [https://www.theatlantic.com/science/archive/2020/05/covid-19-vaccine-skeptics-conspiracies/611998/. Accessed [September 30, 2021].
45. Cohen J. Here's how the U.S. could release a COVID-19 vaccine before the election—And why that scares some. *Science.* 2020. https://doi.org/10.1126/science.abe5150. Published online August 28.
46. Irfan U. Why it's unlikely we'll have a COVID-19 vaccine before election day. *Vox.* 2020; [https://www.vox.com/2020/8/20/21359026/coronavirus-vaccine-COVID-19-when-october-surprise-election-trump Accessed [September 30, 2021].
47. Emanuel EJ, Offit PA. Could trump turn a vaccine into a campaign stunt. *New York Times.* 2020; [https://www.nytimes.com/2020/06/08/opinion/trump-coronavirus-vaccine.html. Accessed [September 30, 2021].
48. Lederman M, Mehiman MJ, Youngner S. Defeat COVID-19 by requiring vaccination for all. It's not un-American, it's patriotic. *USA Today.* 2020; [https://www.usatoday.com/story/opinion/2020/08/06/stop-coronavirus-compulsory-universal-vaccination-column/3289948001/. Accessed [September 30, 2021].
49. Dougherty MB. Vaccine passports are a terrible idea. *Natl Rev.* 2021; [https://www.nationalreview.com/2021/03/vaccine-passports-are-a-terrible-idea/. Accessed [September 30, 2021].

50. Bokemper SE, Huber GA, Gerber AS, James EK, Omer SB. Timing of COVID-19 vaccine approval and endorsement by public figures. *Vaccine*. 2021;39(5):825–829. https://doi.org/10.1016/j.vaccine.2020.12.048.
51. Abbas AH. Politicizing COVID-19 vaccines in the press: A critical discourse analysis. *Int J Semiot Law Published online July*. 2021;10. https://doi.org/10.1007/s11196-021-09857-3.
52. Thigpen CL, Funk C. *Most Americans Expect a COVID-19 Vaccine within a Year; 72% Say they Would Get Vaccinated*. Pew Research Center; 2021 [https://www.pewresearch.org/fact-tank/2020/05/21/most-americans-expect-a-covid-19-vaccine-within-a-year-72-say-they-would-get-vaccinated/. Accessed [September 30, 2021].
53. Tyson A, Johnson C, Funk C. *U.S. Public Now Divided over whether to Get COVID-19 Vaccine*. Pew Research; 2020 [https://www.pewresearch.org/science/2020/09/17/u-s-public-now-divided-over-whether-to-get-covid-19-vaccine/. Accessed [September 30, 2021].
54. Callaghan T, Moghtaderi A, Lueck JA, et al. Correlates and disparities of COVID-19 vaccine hesitancy. *Soc Sci Med*. 2020;272:113638. https://doi.org/10.1016/j.socscimed.2020.113638.
55. Pink SL, Chu J, Druckman J, Rand DG, Willer R. Elite party cues increase vaccination intentions among republicans. *PsyArXiv*. 2021. https://doi.org/10.31234/osf.io/f9jq5.
56. Gauchat G. Politicization of science in the public sphere: A study of public trust in the United States, 1974 to 2010. *Am Sociol Rev*. 2012;77(2):167–187. https://doi.org/10.1177/0003122412438225.
57. Oliver JE, Rahn WM. Rise of the *Trumpenvolk*: Populism in the 2016 election. *Ann Am Acad Pol Soc Sci*. 2016;667(1):189–206. https://doi.org/10.1177/0002716216662639.
58. Motta M. Can a COVID-19 vaccine live up to Americans' expectations? A conjoint analysis of how vaccine characteristics influence vaccination intentions. *Soc Sci Med*. 2021;272:113642. https://doi.org/10.1016/j.socscimed.2020.113642.
59. Pickup M, Stecula D, Van Der Linden C. Novel coronavirus, old partisanship: COVID-19 attitudes and behaviours in the United States and Canada. *Canadian Journal of Political Science/Revue*. 2020 Jun;53(2):357–364.
60. Druckman JN, McGrath MC. The evidence for motivated reasoning in climate change preference formation. *Nature Clim Change*. 2019;9(2):111–119. https://doi.org/10.1038/s41558-018-0360-1.
61. Druckman J, Klar S, Krupnikov Y, Levendusky M, Ryan JB. The political impact of affective polarization: How partisan animus shapes COVID-19 attitudes. *PsyArXiv*. 2020. https://doi.org/10.31234/osf.io/ztgpn.
62. Bolsen T, Palm R. Motivated reasoning and political decision making. In: *Oxford Research Encyclopedia of Politics*. Oxford University Press; 2019. https://doi.org/10.1093/acrefore/9780190228637.013.923.
63. Kunda Z. The case for motivated reasoning. *Psychol Bull*. 1990;108(3):480–498. https://doi.org/10.1037/0033-2909.108.3.480.
64. Sylvester SM. COVID-19 and motivated reasoning: The influence of knowledge on COVID-related policy and health behavior. *Soc Sci Q*. 2021. https://doi.org/10.1111/ssqu.12989. Published online May 25, 2021:ssqu.12989.
65. Bayes R, Druckman JN, Goods A, Molden DC. When and how different motives can drive motivated political reasoning. *Polit Psychol*. 2020;41(5):1031–1052. https://doi.org/10.1111/pops.12663.
66. Kahan DM, Jenkins-Smith H, Braman D. Cultural cognition of scientific consensus. *J Risk Res*. 2011;14(2):147–174. https://doi.org/10.1080/13669877.2010.511246.
67. Douglas KM, Uscinski JE, Sutton RM, et al. Understanding conspiracy theories. *Polit Psychol*. 2019;40(S1):3–35. https://doi.org/10.1111/pops.12568.

68. Enders AM, Uscinski JE, Klofstad C, Stoler J. The different forms of COVID-19 misinformation and their consequences. *HKS Misinfo Review*. 2020. https://doi.org/10.37016/mr-2020-48. Published online November 12.
69. Bolsen T, Palm R, Kingsland JT. Framing the origins of COVID-19. *Sci Commun*. 2020;42(5):562–585. https://doi.org/10.1177/1075547020953603.
70. Flynn DJ, Nyhan B, Reifler J. The nature and origins of misperceptions: Understanding false and unsupported beliefs about politics. *Adv Exp Soc Psychol*. 2017;38:127–150. https://doi.org/10.1111/pops.12394.
71. Miller JM, Saunders KL, Farhart CE. Conspiracy endorsement as motivated reasoning: The moderating roles of political knowledge and trust. *Am J Pol Sci*. 2016;60:824–844.
72. Uscinski JE, Parent JM. *American Conspiracy Theories*. Oxford University Press; 2014. https://doi.org/10.1093/acprof:oso/9780199351800.001.0001.
73. Kupferschmidt K. Evolving threat. *Science*. 2021;373(6557):844–849. https://doi.org/10.1126/science.373.6557.844.
74. World Health Organization. *Tracking SARS-CoV-2 Variants*; 2021 [https://www.who.int/en/activities/tracking-SARS-CoV-2-variants/. Accessed [September 30, 2021].
75. Telenti A, Arvin A, Corey L, et al. After the pandemic: Perspectives on the future trajectory of COVID-19. *Nature*. 2021;596(7873):495–504. https://doi.org/10.1038/s41586-021-03792-w.
76. Rochman ND, Wolf YI, Faure G, Mutz P, Zhang F, Koonin EV. Ongoing global and regional adaptive evolution of SARS-CoV-2. *Proc Natl Acad Sci U S A*. 2021;118(29). https://doi.org/10.1073/pnas.2104241118, e2104241118.
77. Hamel L, Lopes L, Kearney A, Sparks G, Stokes M, Brodie M. KFF Covid-19. *Vaccine Monitor*. 2021; [June 30, 2021. https://www.kff.org/coronavirus-covid-19/poll-finding/kff-covid-19-vaccine-monitor-june-2021/. Accessed [September 30, 2021].
78. Galston WA. What drives GOP resistance to vaccines? *Wall Street Journal*. 2021; [https://www.wsj.com/articles/what-drives-gop-resistance-to-vaccines-11623190392. Accessed [September 30, 2021].
79. Cohan A. *Timeline: Changes to CDC Mask Guidelines since the Pandemic Began*. Boston Herald; 2021 [https://www.bostonherald.com/2021/07/27/timeline-changes-to-cdc-mask-guidelines-since-the-pandemic-began/. Accesssed [September 30, 2021].
80. Ellis R. Ex-surgeon general: CDC mask guidance may be 'wrong'. *WebMD News Brief*. 2021; [https://www.webmd.com/lung/news/20210720/ex-surgeon-general- -says-cdc-mask-guidance-may-be-wrong. Accessed [September 30, 2021].
81. Aratani L. How did face masks become a political issue in America. *The Guardian*. 2020; [https://www.theguardian.com/world/2020/jun/29/face-masks-us-politics-coronavirus. Accessed [September 30, 2021].
82. Lombardo C. In Florida, Texas and Arizona, defiant school leaders are sticking with mask mandates. *GPB News*. 2021; [https://www.gpb.org/news/back-school-live-updates/2021/08/18/in-florida-texas-and-arizona-defiant-school-leaders-are. Accessed [September 30, 2021].
83. Stirgus E. Georgia university system leader defends COVID prevention methods. *Atlanta Journal Constitution*. 2021; [https://www.ajc.com/education/georgia-university-system-leader-defends-covid-prevention-methods/R3TGI55Q6RCDDO3CDAYAXWDSGA/. Accessed [September 30, 2021].
84. Camera L. GOP governors double down on orders barring mask mandates in schools. *US News*. 2021; [https://www.usnews.com/news/education-news/articles/2021-08-10/gop-governors-double-down-on-orders-barring-mask-mandates-in-schools. Accessed [September 30, 2021].
85. Morris K. Republicans explode with fury over Biden vaccine mandate: 'Absolutely unconstitutional.'. *Fox News*. 2021; [https://www.foxnews.com/politics/republicans-respond-to-vaccine-mandates-issued-by-biden-absolutely-unconstitutional. Accessed [September 30, 2021].

86. Blake A. *The GOP's Halting, Uneven Journey toward Becoming the Anti-Vaccine Mandate Party*. Washington Post; 2021 [https://www.washingtonpost.com/politics/2021/09/11/gops-halting-journey-opposing-vaccine-mandates/. Accessed [September 30, 2021].
87. Stacey K. Republicans threaten to block Biden's "dictatorial" Covid vaccine mandate. *Financial Times*. 2021; [https://www.ft.com/content/144be751-f69a-4d20-871c-c191c2b2d4f9. Accessed [September 30, 2021].
88. Capehart J. DNC chair: GOP reaction to Biden vaccine mandate is "unhinged.". *MSNBC interview*. 2021; [https://www.youtube.com/watch?v=J0XYUfQhhIw. Accessed [September 30, 2021].
89. Green J, Druckman J, Baum M, et al. *Using general messages to persuade on a politicized scientific issue*. Working Paper; 2021.
90. Lewandowsky S, Gignac GE, Vaughan S. The pivotal role of perceived scientific consensus in acceptance of science. *Nat Clim Change*. 2013;3(4):399–404. https://doi.org/10.1038/nclimate1720.
91. van der Linden S. The gateway belief model (GBM): A review and research agenda for communicating the scientific consensus on climate change. *Curr Opin Psychol*. 2021. https://doi.org/10.1016/j.copsyc.2021.01.005.
92. van der Linden SL, Clarke CE, Maibach EW. Highlighting consensus among medical scientists increases public support for vaccines: Evidence from a randomized experiment. *BMC Public Health*. 2015;15(1):1207. https://bmcpublichealth.biomedcentral.com/articles/10.1186/s12889-015-2541-4.
93. Motta M, Sylvester S, Callaghan T, Lunz-Trujillo K. Encouraging COVID-19 vaccine uptake through effective health communication. *Front Polit Sci*. 2021;3:630133. https://doi.org/10.3389/fpos.2021.630133.
94. Bolsen T, Druckman JN. Counteracting the politicization of science. *J Commun*. 2015;65(5):745–769. https://doi.org/10.1111/jcom.12171.
95. Bolsen T, Druckman JN. Do partisanship and politicization undermine the impact of a scientific consensus message about climate change? *Group Process Intergr Relat*. 2018;21(3):389–402. https://doi.org/10.1177/1368430217737855.
96. Cook J, Lewandowsky S, Ecker UKH. Neutralizing misinformation through inoculation: Exposing misleading argumentation techniques reduces their influence. *PLoS ONE*. 2017;12(5):e0175799. https://doi.org/10.1371/journal.pone.0175799.
97. van der Linden S, Leiserowitz A, Rosenthal S, Maibach E. Inoculating the public against misinformation about climate change. *Global Challenges*. 2017;1(2):1600008.
98. Iyengar S, Massey DS. Scientific communication in a post-truth society. *Proc Natl Acad Sci U S A*. 2019;116(16):7656–7661. https://doi.org/10.1073/pnas.1805868115.
99. Lazer DMJ, Baum MA, Benkler Y, et al. The science of fake news. *Science*. 2018;359(6380):1094–1096. https://doi.org/10.1126/science.aao2998.
100. Levy J, Bayes R, Bolsen T, Druckman JN. Science and the politics of misinformation. In: Tumber H, Waisbord S, eds. *The Routledge Companion to Media Disinformation and Populism*. 1st ed. Routledge; 2021. https://doi.org/10.4324/9781003004431.
101. Nyhan B, Reifler J, Richey S, Freed GL. Effective messages in vaccine promotion: A randomized trial. *Pediatrics*. 2014;133(4):e835–e842. https://doi.org/10.1542/peds.2013-2365.
102. Nyhan B, Reifler J. When corrections fail: The persistence of political misperceptions. *Polit Behav*. 2010;32(2):303–330. https://doi.org/10.1007/s11109-010-9112-2.
103. Carey JM, Guess AM, Loewen PJ, Merkley E, Nyhan B, Phillips JB, Reifler J. The ephemeral effects of fact-checks on COVID-19 misperceptions: Evidence from the US, Great Britain, and Canada. September 30, 2021. Working paper.
104. Van Bavel JJ, Baicker K, Boggio P, et al. Using social and Behavioural science to support COVID-19 pandemic response. *PsyArXiv*. 2020. https://doi.org/10.31234/osf.io/y38m9.

105. Druckman JN, Ognyanova K, Baum MA, et al. The role of race, religion, and partisanship in misperceptions about COVID-19. *Group Process Intergr Relat.* 2021;24 (4):638–657. https://doi.org/10.1177/1368430220985912.
106. Resnik DB. *Playing Politics with Science: Balancing Scientific Independence and Government Oversight.* Oxford University Press on Demand; 2009.
107. Burke J. *The Day the Universe Changed.* Boston: MA: Little and Brown; 1995.
108. Kuhn T. *The Structure of Scientific Revolutions.* 2nd ed. Chicago, IL: University of Chicago Press; 1970.
109. Campbell N, Roper-Fetter K, Yoder M. *Principles of Risk Communication: A Guide to Communicating with Socially Vulnerable Populations across the Disaster Lifecycle.* Boulder, CO: Natural Hazards Center, University of Colorado Boulder; 2021.
110. Norton A, Olliaro P, Sigfrid L, et al. Long COVID: Tackling a multifaceted condition requires a multidisciplinary approach. *Lancet Infect Dis.* 2021;21(5):601–602. https://doi.org/10.1016/S1473-3099(21)00043-8.
111. Rutjens BT, van der Linden S, van der Lee R. Science skepticism in times of COVID-19. *Group Process Intergr Relat.* 2021;24(2):276–283. https://doi.org/10.1177/1368430220981415.

CHAPTER SIX

Effects of politicized media coverage: Experimental evidence from the HPV vaccine and COVID-19

Erika Franklin Fowler[a], Rebekah H. Nagler[b], Darshana Banka[a], and Sarah E. Gollust[b,*]

[a]Wesleyan University, Middletown, CT, United States
[b]University of Minnesota, Minneapolis, MN, United States
*Corresponding author: e-mail address: sgollust@umn.edu

Contents

1. The science of guidelines and updating health recommendations 103
2. Concerns over politicization, dimensions of the concept, and strategies for mitigation 104
   2.1 Politicization concerns and consequences 104
   2.2 Concerns about potential carryover effects of politicization 105
   2.3 Dimensions of politicization 105
   2.4 Messaging strategies to mitigate politicization 106
3. Study objectives 107
4. Data and methods 109
   4.1 Study design 109
   4.2 Measures 112
   4.3 Analytic approach 114
5. Results 115
   5.1 Negative emotional reactions to politicization and inoculation 115
   5.2 Domain specific policy support responses to politicization and inoculation 118
   5.3 Confusion and backlash responses to politicization and inoculation 119
   5.4 Potential carryover effects of politicization and inoculation 122
6. Discussion 125
Appendix 128
   A.1 Appendix: Demographic sample 128
   A.2 Appendix: Treatment language 129
References 131

Progress in Molecular Biology and Translational Science, Volume 188
ISSN 1877-1173
https://doi.org/10.1016/bs.pmbts.2021.11.008

Abstract

Although concerns about politicization of health and science are not new, the COVID-19 pandemic has amplified attention to how political disagreement over scientific guidelines and recommendations might influence attitudes and behaviors about the health topics in question and might even spill or carry over to affect other attitudes important to public health. The literature employs differing definitions of politicization—at times referring to controversy in the public sphere, at others referring to the exploitation of the uncertainty inherent in science, and at still others referring to whether the issue enters political discourse—all of which are viewed as distinct dimensions by the public. What is not known is how these different aspects of politicization influence public attitudes about the health topics and or broader attitudes about scientific guidelines, and—assuming adverse effects—what strategies might be effective at mitigating the consequences. This paper draws on a survey experiment of 3012 U.S. respondents fielded in summer 2020 that was designed as a pilot study to assess the effects of different dimensions of politicization. Findings do not suggest that one type of politicization is necessarily more pernicious than the others. In fact, all types of politicization increased negative emotional responses and confusion, both with respect to the health topic in question (HPV vaccine and COVID-19) but also on other domains, although opinions about policy were unaffected. The findings also suggest that inoculation may have potential as a messaging strategy for blunting the adverse effects of exposure to politicization.

Although scholarly concerns about politicization of health and science are not new,[1–6] the COVID-19 pandemic has increased attention to how political disagreement over scientific guidelines and recommendations might influence the public's attitudes and behaviors, both about the health topics in question[7–15] as well as other attitudes important to public health.[16] As prior work has acknowledged, the literature employs differing definitions of politicization—at times referring to controversy in the public sphere,[17] at others referring to the exploitation of the uncertainty inherent in science,[6] and at still others referring to the extent to which the issue enters political discourse.[18] In previous work, we have found that the public perceives these distinct dimensions and that these dimensions have different associations with policy attitudes.[19,20] What is not known is the causal impact of these different aspects of politicization on public attitudes about the topics in question or more general attitudes about scientific guidelines and recommendations. And should there be adverse effects of different dimensions of politicized messages about health topics, there is insufficient research on what strategies might be most effective at mitigating each type.

In this paper, we draw on a survey experiment that was designed as a pilot study to assess the effects of different dimensions of politicization (i.e., emphasizing controversy, scientific uncertainty, and prevalent partisan political cues). We randomly assigned participants to a topic (HPV vaccine or COVID-19) and then to receive either general information about updated scientific recommendations on those topics or to information about those recommendations that includes differing forms of politicization about the topic. We also tested whether one specific type of messaging strategy—an inoculation message tailored to each dimension that warns of people seeking to politicize health and science issues, delivered in advance of the politicization message treatments—can blunt the effects of politicization. The findings do not suggest that one type of politicization is necessarily more pernicious than the others. In fact, all types of politicization increased negative emotional responses and confusion, both with respect to the health topic in question but also on related domains, although policy support remained unchanged. The results also suggest that inoculation may sometimes blunt the adverse effects of exposure to politicization, although testing additional strategies is needed, given the increasing frequency of politically-charged and controversial health messaging in the media ecosystem.

1. The science of guidelines and updating health recommendations

The scientific enterprise depends upon the accumulation of knowledge, which is acquired through testing and retesting different hypotheses and systematic syntheses of existing studies. As new technologies and methods are developed, the evidence base guiding procedures and recommendations for whom, when and how new innovations should be implemented (or de-implemented[21,22]) and used grows. Of course, who interprets the evidence and how recommendations and guidelines are established is a social process that can be shaped by politics. In the context of health care, although the United States has been comparatively slow to implement evidence-based medicine recommendations into public policy in large part due to politics[23] and a lack of public understanding,[24–27] guidelines for medical care recommendations have been long been established by professional organizations and sometimes by panels of independent medical experts who are convened for specific reasons. For example, the Advisory Committee on Immunization Practices (ACIP) was established in 1964 by the Surgeon General to provide independent, outside advice to the Centers for Disease Control, and the U.S. Preventive Services Task Force was created in 1984 to examine evidence and develop recommendations for primary and preventive care.

As research accumulates and evolves, the recommendations and guidelines experts follow should be expected to change in accordance with the body of evidence, and yet the way in which guidelines are updated and described in public discourse can influence acceptance of the recommendations. More specifically, the public is less trusting of expert panels described to be government sponsored[24] and less likely to support politicians who challenge medical experts on the appropriateness of particular treatments even if the evidence supports the politician's argument.[23] While some recommendations like the push to move colorectal cancer screening earlier (i.e., from 50 to 45) have largely been accepted without public pushback or scrutiny, others have been sparked with controversy and concerns over politicization.[28]

2. Concerns over politicization, dimensions of the concept, and strategies for mitigation

2.1 Politicization concerns and consequences

Although some scholars argue that science is inherently political and uncertain and therefore politicization will always be a feature of public communication about these issues—especially in the realm of public health topics[3,29]—others argue that there are significant concerns related to the politicization of health and science issues, especially when it is a prominent feature of public discourse surrounding scientific recommendations. In particular, prior research has articulated concerns and/or documented evidence demonstrating politicization leads to a host of adverse responses including: a lack of consensus over proper policy action and support for science,[2] decreased trust in the medical profession and doctors[18] and in science overall,[1] along with jeopardizing funding for basic scientific research.[4] These warnings led to a growing body of work attempting to mitigate the adverse consequences of politicization,[30,31] even before the COVID-19 pandemic served to highlight just how deadly the consequences of politicized science and polarized partisan reactions really are.[7,14,32,33]

Several earlier studies about the adverse influence of politicization have generated evidence that it increases negative emotional reactions, heightens doubt about scientific claims, and decreases the uptake of beneficial technologies or policies in the public interest. For example, prior work has shown that politicization messaging increases anxiety and perceptions of threat.[6] Studies have shown that news coverage providing equal coverage to opposing views of scientific claims, whether on climate science or other health issues, results in increased uncertainty about the veracity of scientific claims.[34] In other cases, politicized framing of news can result in reductions of support for nuclear power,[35] hydraulic fracturing and carbon nanotubes,[6] and for requiring the HPV vaccine.[36] The politicized nature of the discussion surrounding the Affordable Care Act has even contributed to differential uptake of health insurance based on enrollees' party identification, suggesting that partisan reactions can have real impacts on health.[37–39]

2.2 Concerns about potential carryover effects of politicization

Exposure to politicized messages might produce effects not only on emotional responses to and public attitudes about the health topics in question, but also on more general attitudes about scientific guidelines and recommendations. Such spillover or "carryover" effects have been observed in response to exposure to health-related conflict and controversy. For example, exposure to conflicting information about one health topic (e.g., mammography) has been shown to influence cognitive and behavioral responses about other health topics (e.g., ambivalence not only to mammography but also other types of cancer screening[40]); such exposure also can affect cognitions that are not topic-specific (e.g., attitudes toward health research in general).[41,42] There is even some evidence that these cognitive responses to conflict can decrease receptivity to subsequent unrelated health messages, including those about behaviors for which there is broad scientific consensus (e.g., fruit and vegetable consumption, physical activity).[16] This body of research suggests that politicized health content—especially when characterized by controversy—could carry over to other health topics, thus jeopardizing a much broader set of preventive medicine and public health recommendations whose effects might be felt long after the COVID-19 pandemic.[16]

2.3 Dimensions of politicization

Although numerous pieces of scholarship have examined the causes and consequences of politicization of health and science,[1,3,6,7,10,13,18,35,43] how politicization is defined and operationalized (if at all) differs from study to study. In particular, prior work[19,20] has identified at least three dimensions to how politicization in media messaging is referenced in the literature: as disagreement and controversial discussion of science in the public sphere[17]; as amplification and exploitation of the inherent uncertainty of science in order to cast doubt on guidance[6]; and as the inclusion and incorporation of partisan, political conflict in discussions of health and science topics.[18,36] What is commonly held between all of these articulations is the notion that politicization occurs in public discussion over health and science recommendations, and that the communication surrounding these issues is contested. Although disagreement and controversy is one specific dimension or operationalization of politicization in media, both of the others (amplifying scientific uncertainty and incorporating partisan conflict) add additional layers of messaging on top of the emphasized feature of controversy. In other words, scientific uncertainty and partisan dimensions of politicization might be thought of as additional features layered atop controversial depictions of issues.

Does the type of politicization matter for the effects we see on emotions, public attitudes and behaviors? Because most of the literature utilizes one of the dimensions (if it does define one at all), we know little about how the different types of dimensions might differently influence public reactions. Therefore, this study attempts to test the variety of different effects that politicization has been shown to have on affective, cognitive and policy responses and to assess whether any one type is particularly problematic compared to the others and to a control condition that describes the topic in question without politicized dimensions. Because there is no prior work comparing the effects of different types of politicization, we view our work here as primarily exploratory, but to the extent that both of the uncertainty and political dimensions build on controversy, one might expect that the effects could be stronger for either of those types as compared to controversy alone.

2.4 Messaging strategies to mitigate politicization

Given the growing literature on the adverse effects of politicized messaging in the public sphere, there has also been burgeoning work on strategies to counteract it.[6,30,31,44] A range of possible tactics to mitigate politicization have been explored and include warnings in advance or "inoculation" messages,[6,45,46] corrections after prior exposure,[6,47] visual imagery,[30] self-affirmation,[44] and the provision of specific information on causes of scientific findings.[31] Although many of these strategies may merit examination with respect to individual dimensions of politicization, we focus here on the first: inoculation messages or warnings in advance of politicized messages. Inoculation has been conceptualized as a persuasion strategy that includes a weakened or refuted version of a message (that is to say, a weaker form of a later message, to apply the vaccination metaphor) that helps individuals to develop psychological resistance to a subsequent communication.[48,49] In a previous study related to public health messaging (i.e., soda taxes), this was conceptualized as a warning about the motives of a communicator and preemptive refutation of that messenger's argument.[45] In work specifically on politicization (defined as strategic emphasis of scientific uncertainty), it has been conceptualized as a warning that scientific consensus exists and future politicization should be disregarded.[6] The key feature of an inoculation-style message is that it is delivered in advance of another message, in order to increase the likelihood that people resist being persuaded by (or adversely affected by, in the case of the current study) a subsequent communication. To our knowledge, such a messaging strategy has not been studied in the context of comparing its effectiveness in light of differing types of politicized communication, but we expect inoculation to help blunt the effects of politicization.

3. Study objectives

While the dire nature of the consequences of politicization generally may have been clarified with the onset of the pandemic and the concomitant of polarized responses among the public,[7,14] what is not yet clear is whether the type of politicization matters. As described earlier, prior work differs in the definition and operationalization of politicization, using at least three distinct dimensions of the concept.[19] In particular, it is not clear whether disagreement and controversy itself is the most damaging or whether exploiting the uncertainty of science to create controversy vs the intrusion of partisan cues in controversy over science are worse. As such, we designed a pilot study in the context of HPV vaccine recommendations to determine whether we could uncover differential effects from each of the three distinct types of politicization across a range of outcomes, including affective responses, cognitive responses, and support for policies. In addition, because we conducted the study during the early stages of the pandemic, which was fundamentally characterized by extensive politicization in the United States, we also decided to examine the extent to which exposure to additional politicization of mask-wearing recommendations surrounding COVID-19—operationalized as the presence of all three dimensions, as by July 2020 all three were already indistinguishable in COVID-19 discourse—would influence the same types of outcomes. Thus our study examines the consequences of politicization messaging across two distinct cases, the HPV vaccine and COVID-19, and we examine outcomes that encompass both domain-specific attitudes (i.e., emotional arousal, confusion, backlash, and policy support related to the assigned topic), as well as potential carryover attitudes about vaccination more generally (i.e., confusion on other issues as well as policy support for other issues).

Following the prior literature, we expect that all types of politicization conditions will increase negative emotions, decrease policy support and increase confusion and backlash regarding the specific domain, and that these effects might carry over to related domains (i.e., increase carryover confusion and decrease support for policies beyond the specific domain). Is one type of politicization in messaging worse than another? As mentioned above, we do not have strong expectations about which dimension of politicization may be the most harmful. To the extent that the uncertainty and partisan domains add on additional layers to controversy, it might be reasonable to suggest that both could amplify the adverse effects compared to controversy, but our goal is simply to explore the effects of each type. Therefore, we frame the comparison of different dimensions as a research question: Are the dimensions of politicized messaging different from one another in their influence?

Finally, as noted above, there has been little research that examines strategies to help the public combat or mitigate any adverse effects of different types of politicized messaging. Thus, we also included an initial test of an inoculation strategy to preemptively help people consider that politicization might be a strategic construction of health issues, and that they do not have to be persuaded by it. As mentioned above, we hypothesize that inoculation will help to mitigate potential adverse effects—that is, relative to exposure to politicized messaging alone, it will lower negative emotional reactions, increase policy support and decrease confusion on the domain in question and will increase policy support and decrease confusion on potential carryover topics—of politicized communication. One important caveat to these expectations is that pretreatment, or citizens' prior exposure to information similar to any treatment language, can affect subsequent responses to experimental conditions.[50] Because politicization was already baked into much of the communication surrounding COVID-19 in particular, there is reason to believe that people may not be as responsive to an additional dose of politicization, as they may have already adjusted their attitudes and beliefs in response to earlier messaging.

We discuss each of the features of the study design below.

4. Data and methods

4.1 Study design

In this analysis, we draw on a survey experiment of 3012 respondents,[a] which was fielded online between June 30 and July 6, 2020, by Dynata on a national sample of U.S. adults. In particular, the sample was designed to be representative of U.S. adults 18 years or older with recruitment designed to achieve U.S. Census estimates on two characteristics: age and gender. Participants were randomly assigned to a topic first (either HPV vaccine recommendations or mask-wearing recommendations surrounding the spread of COVID-19) and then to either receiving a news article with information about updated scientific recommendations on that topic (hereafter called base level information or base information) or to a news article with the same base information about recommendations that also included content containing differing forms of politicization (see Table 1 for headlines and inoculations by topic and see the Appendix for the complete treatment language).

The base information for both the HPV vaccine and for the COVID-19 condition contained news about updated recommendations in light of evidence (or the accumulation of additional data as would be expected in the normal progression of science). More specifically, the HPV vaccine condition (labeled as an August 2019 news article) stated that the Advisory Committee on Immunization Practices (ACIP) recommended that men and women up to age 26 receive a "catch-up" vaccine if they missed the shots in pre-adolescence, whereas the COVID-19 condition (labeled as a June 2020 news article) provided information about the updated recommendations from the CDC advising that people wear masks in public. In all politicization treatments, we modified the headline of the news article to emphasize the specific aspect of politicization and added both a paragraph describing heated debate along with a quote highlighting the politicization to emphasize the disagreement.

[a] The data we present here ($N = 3012$) is a subset of a larger experiment ($N = 6027$) that included an additional topic (mammography) along with a pure control condition that did not see a base message at all. Because we have evidence that the public does not perceive mammography to be politicized at all, we chose only to focus here on HPV and COVID-19 because they are more similar but still represent different ranges of perceived politicization. Further, because we care most about how politicization treatments differ from base level information about scientific recommendations, we do not utilize comparisons to the pure control condition in this analysis.

Table 1 Headlines and inoculation treatments for each of the experimental study conditions.

	HPV vaccine headlines	COVID-19 headlines	Inoculation
Base	Panel to Expand HPV Vaccine Recommendations	CDC Updates Recommendations on Mask Wearing	
+Controversy	Controversy Erupts Over New HPV Vaccine Recommendations		…people…try to make health recommendations seem more controversial than they really are…. They create drama to draw attention away from the large amount of agreement…
+Controversy +Uncertainty	Advocates Cast Doubt on Scientific Evidence Behind New HPV Vaccine Recommendations as Controversy Erupts		…people…try to make health recommendations seem more controversial than they really are, saying that the science is more uncertain than it really is…. They create drama and exaggerate uncertainty to draw attention away from the large amount of scientific agreement…
+Controversy +Political	Political Pushback to New HPV Vaccine Recommendations as Controversy Erupts		…politicians…try to make health recommendations seem more controversial than they really are…. They create drama for political gain to draw attention away from the large amount of bipartisan agreement…
All Politicized Elements		Controversy and Political Pushback Over CDC Mask Wearing Recommendations in Light of Doubts About Scientific Evidence	…politicians or other people…try to make health recommendations seem more controversial than they really are, saying that the science is more uncertain than it really is or that there is more division between Republicans and Democrats than there really is…. They create drama and exaggerate uncertainty to draw attention away from the large amount of scientific and bipartisan agreement…

Note: News article treatment language varied between the two topics. Each of the politicized conditions added additional language (in addition to the headline changes listed in the table) to the base information to help emphasize the particular treatment which included discussion of heated debate and disagreement and a quote.

Specific details of the treatments varied by topic. In the case of the HPV vaccine, we tested three separate politicization conditions: (1) one that exposes participants to disagreement and controversy surrounding the updated recommendations, as well as a "local advocate" summing up the concerns (the Base + Controversy condition, hereafter controversy for short), (2) one that adds explicit discussion of the uncertainty of the science underpinning the recommendations, as well as a "local advocate" questioning the science (the Base + Controversy + Uncertainty condition, hereafter uncertainty for short), and (3) one that adds political discourse, specifically a "longtime politician" weighing in, questioning, and highlighting the controversy over the recommendations (the Base + Controversy + Political condition, hereafter political for short). In the case of COVID-19, we were fielding the study during the early stage of the pandemic when attention to news about coronavirus was especially high and politicization of health recommendations had already begun.[7] Therefore, we decided that we could not reliably assess the different dimensions of politicization given the interference of pretreatment messaging, which likely varied in ways we could not adequately control posthoc. Therefore, we only tested one combined politicization condition that contained all three elements (controversy, uncertainty, and political discourse, each part of the broad public discourse surrounding masking as a prevention strategy) as a way of understanding how the politicization emphasis in news coverage affected responses. In the COVID-19 politicization treatment, similar to the HPV vaccine, we added a paragraph on heated debate along with a quote that further emphasized it along with the headline. The headlines for each of the six conditions are listed in the middle two columns of Table 1 (and for the full language, see the Appendix).

Finally, we randomly assigned half of respondents who were not assigned to the base condition to view an inoculation treatment in advance of receiving the news media article. The initial prompt for inoculation treatments said, "On the next page, you will see a statement followed by a recent news article published in a local paper. Please take a moment to read it and you will be asked a few questions about it afterward." As shown in the far right column of Table 1, the inoculation language warned of actors emphasizing controversy, controversy and uncertainty, or controversy and political influence (depending on the assigned politicization dimension for HPV vaccine; or all of the above elements for COVID-19); the message strategy here was to help respondents refute or dismiss the impact of these politicized communication dimensions when they see them in the news article. Finally, Table 2 provides an overview of the sample size for each of the experimental design conditions.

Table 2 Experimental conditions and sample size by topic.

	HPV vaccine		COVID-19	
	No Inoc.	+ Inoc.	No Inoc.	+ Inoc.
Base	296		303	
+Controversy	305	308		
+Controversy +Uncertainty	286	303		
+Controversy +Political	300	296		
All Politicized Elements			302	313
Total $N = 3012$	2094		918	

4.2 Measures

As described above, based on previous research examining the effects of exposure to politicized messages and conflicting information (a related, but not identical, concern as politicization), we conceptualized five sets of outcome measures, adapted from previous studies[6,16,20,36,40,51]: negative emotional responses; support for the policy action in the specific domain; confusion or perceived ambiguity about the specific topic, backlash or negative beliefs toward the specific topic, and a set of potential carryover effects—in this case, general confusion about broader topics than those featured in the experimental treatment, as well as vaccine-related attitudes distinct from the health topic in question.

4.2.1 Negative emotional responses

We measured emotional reactions by asking respondents to indicate how they feel after having read the news article with respect to five different emotions: interested, frustrated, surprised, annoyed and distressed. Given our a priori interests in negative emotional responses and consistent with past research,[16,40] we focused on the three negatively-valenced items (frustrated, annoyed, and distressed). Response options ranged from "very slightly or not at all" (1) to "extremely" (5). The distressed item has mean of 2.05 and a standard deviation of 1.19. As the frustrated and annoyed items scaled reliably (reliability coefficient = 0.82), we combined them into an irritation index (mean = 2.16; SD = 1.14).

4.2.2 Support for domain specific policy action

With respect to domain-specific policy support, we utilized a question from prior literature in the HPV vaccine context[36]: "How much do you support or oppose a state law requiring all children in middle school to get the HPV vaccine with opt-outs allowed as determined by your state?" Response options ranged from (1) strongly oppose to (7) strongly support (mean = 4.44; SD = 1.82). In the case of COVID-19, we asked two questions about how much the participant supports or opposes "local governments requiring people to wear masks in public" and "businesses (e.g., stores, restaurants) requiring people to wear masks in their establishments." Both measures ranged from (1) strongly oppose to (7) strongly support. While we had originally conceived that the public may differ in their attitudes toward government mandates versus mandates in the private sector, because the alpha scale reliability coefficient is 0.91, we created one index of the two, which also ranges from 1 to 7 (mean 5.91; SD = 1.72).

4.2.3 Confusion and backlash

To assess HPV vaccine-related confusion, we adapted two items from past research[16,40]: "I am not sure whether getting an HPV vaccine is good or bad for a young person's health" and "I find information about getting an HPV vaccine to be confusing." We combined them into one index, which had a reliability coefficient of 0.77. Response options ranged from (1) strongly disagree to (7) strongly disagree (mean = 3.66; SD = 1.60). Similarly, we adapted two items to assess COVID-19-related mask-wearing confusion: "I am not sure whether mask wearing is an effective or ineffective strategy for preventing COVID-19 (coronavirus) spread" and "I find information about mask wearing effectiveness to be confusing." The scale reliability for the index was 0.91 with response options from (1) strongly disagree to (7) strongly agree (mean = 3.26; SD = 1.62).

To assess HPV vaccine-related backlash, we adapted two items from past work[16,40]: "Scientists really don't know whether getting an HPV vaccine is good or bad for a young person's health" and "Recommendations about getting an HPV vaccine should be taken with a grain of salt" into one index (reliability coefficient 0.81). Response options ranged from (1) strongly disagree to (7) strongly agree (mean = 3.56; SD = 1.59). Two similar items were used to assess COVID-19-related mask-wearing backlash: "Scientists really don't know whether mask wearing is an effective or ineffective strategy for preventing COVID-19 (coronavirus) spread" and "Recommendations about wearing a mask should be taken with a grain of salt," which we combined (reliability coefficient 0.81). Response options also ranged from (1) strongly disagree to (7) strongly agree (mean 3.21; SD = 1.81).

4.2.4 Potential carryover effects

We assessed potential carryover effects in two ways. First, given the context of the HPV vaccine topic, we assessed generalized confusion about childhood vaccinations: "I am not sure whether getting childhood vaccinations is good or bad for children's health." We asked a similar item in the context of COVID-19 to assess broader confusion than just about masks as mitigation: "I am not sure whether there are effective strategies for preventing COVID-19 (coronavirus) spread" (both items had response options from (1) strongly disagree to (7) strongly disagree; HPV vaccine mean = 3.17; SD = 1.79, and COVID-19 mean = 3.52; SD = 1.92). Second, we examined whether there were potential carryover effects on support for immunization program requirements, adapting a set of items from past work.[36] In the HPV vaccine context, this included a five item index of responses to: "Required childhood vaccinations important for protecting the American public from disease," "Government should have authority to require vaccines," "Vaccines protect children's health," "Vaccines protect adults' health," and "Vaccines are safe." Three of these items were used to create an index to measure support for adult immunization programs in the COVID-19 context: "Government should have the authority to require vaccines," "Vaccines protect adults' health," and "Vaccines are safe." Response options for each ranged from (1) strongly disagree to (7) strongly agree. The scale reliability for the general childhood immunization index was 0.92 (mean = 5.30; SD = 1.40) and it was 0.84 for the general adult immunization index (mean = 5.06; SD = 1.54).

4.3 Analytic approach

We used linear regression modeling to compare the politicization conditions to the base condition for each topic and difference of means *t*-tests to assess whether the inoculation conditions decrease adverse reactions compared to the politicization treatments without inoculation messaging. We report one-tailed tests for our directional expectations (adverse effects of exposure to politicization relative to the base conditions and protective effects of inoculation relative to politicized messaging alone) but two-tailed tests for the comparison of each politicized dimension compared to each other in the HPV vaccine case. All analyses were conducted with Stata SE, version 14.2.

5. Results

Is one particular type of politicization more damaging than the others? Or do they all operate similarly with respect to their effect on negative emotions, policy support, confusion and backlash, and carryover responses? And does inoculation provide protection against adverse effects of different dimensions of politicization?

5.1 Negative emotional reactions to politicization and inoculation

We begin with an assessment of negative affective responses, which are perhaps most proximal to the treatment. Prior work suggests that exposure to conflict and politicization (defined as uncertainty) arouses negative emotions.[6,16,40] We find that all dimensions of politicization increase negative emotional reactions. As shown in the left panel of Fig. 1, all three types of politicization in the HPV vaccine case increase respondent irritation compared to those only exposed to base information, although there is no meaningful difference between the three types. In addition, respondents exposed to all three types of politicization also report more distress than those exposed to only the base information (see the right panel of Fig. 1); however, only the controversy and political dimensions distress scores are statistically significantly different from the base treatment ($P=0.026$ and $P=0.048$, respectively). In comparing the three conditions to each other, however, we cannot distinguish between the effects of controversy, uncertainty and political, suggesting that none are particularly worse than the other.

Turning to emotional reactions to politicization in the case of COVID-19, we find strong evidence for both an increase in irritation and an increase in distress for respondents exposed to all three combined elements of politicization, as shown in Fig. 2. There are two noteworthy observations regarding the difference between responses to the HPV vaccine and the COVID-19 cases. First, respondents reported greater negative affect (as reflected in both irritation and distress) after simply reading the updated guidelines regarding masks (the base treatment) than they did after reading politicized versions of the HPV vaccine treatments: The mean values are 2.2 and 2.2, respectively, for irritation and distress for the COVID-19 base treatment, compared to values that range from 2.1 to 2.2 and 2.0 to 2.1, respectively, for the HPV vaccine politicized treatments. Second, the negative emotional reaction to politicization

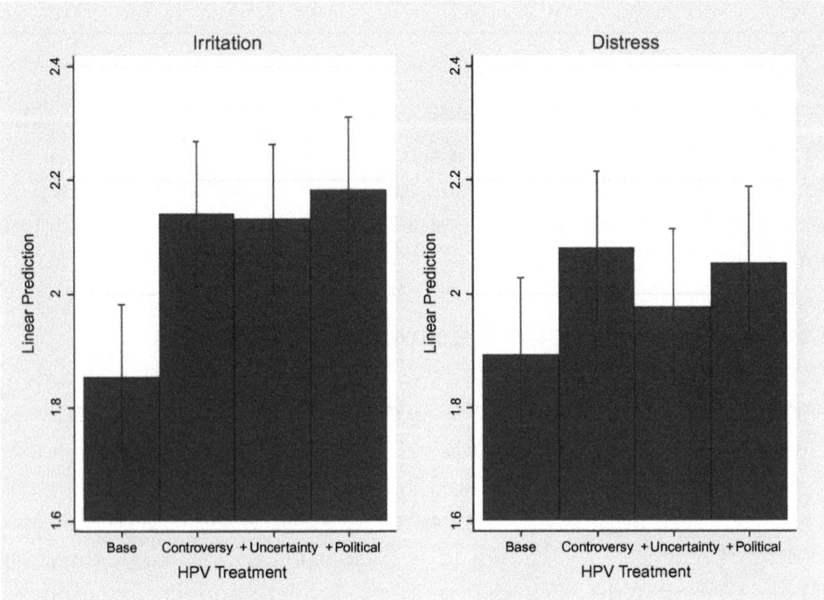

Fig. 1 Effects of politicization treatments on negative emotional reactions (HPV vaccine).

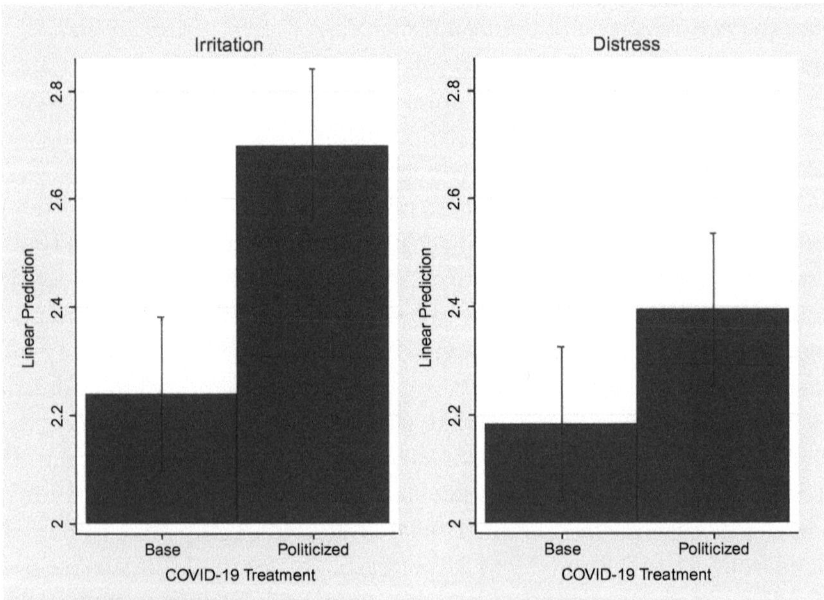

Fig. 2 Effects of Politicization treatments on negative emotional reactions (COVID-19).

(operationalized through all three dimensions) was much larger on both irritation and distress (mean values of 2.7 and 2.4) than it was for any one of the dimensions with respect to the HPV vaccine (all mean negative emotional responses were below 2.2).

Can we successfully blunt the negative emotional effects of politicization with a warning prior to exposure? Table 3 provides mixed and unexpected evidence. More specifically, in comparing the respondents' emotional reactions in the treatment conditions (without inoculation) to those who received inoculation warnings prior to treatment, we find in every case except one (the controversy treatment's effect on distress) that exposure to a warning intended to inoculate respondents against the effect of politicization actually *increased* negative responses. It is important to acknowledge that most of the mean differences between those receiving inoculation and those who did not receive them are not statistically significant; however, in both COVID comparisons, we find a meaningful difference in irritation and in distress above the levels reported by respondents in the politicization condition alone. Furthermore, posthoc *t*-tests between the conditions

Table 3 *T*-test comparisons of inoculation warnings on negative emotions compared to politicization conditions alone.

		Irritation			Distress		
		Mean	SE	p	Mean	SE	p
HPV Vaccine Conditions	*Controversy*			0.92			0.21
	No Inoc.	2.14	0.065		2.08	0.070	
	+ Inoc.	2.27	0.066		2.00	0.067	
	+Uncertainty			0.78			0.92
	No Inoc.	2.13	0.065		1.98	0.068	
	+ Inoc.	2.20	0.063		2.11	0.066	
	+Political			0.92			0.97
	No Inoc.	2.18	0.070		2.05	0.070	
	+ Inoc.	2.32	0.066		2.24	0.073	
COVID-19	*Politicized*			0.99			0.99
	No Inoc.	2.70	0.074		2.39	0.074	
	+ Inoc.	2.95	0.073		2.65	0.077	

Note: One-tailed tests that inoculations are lower than regular treatment cells are reported.

(not shown in in Table 3) with and without inoculation actually suggest that three of the inoculation conditions were statistically higher in negative affect compared to respondents assigned to conditions without inoculation (the political with inoculation condition for the HPV vaccine treatment on distress and the combined politicization with inoculation COVID treatment conditions for both irritation and distress) and three more were marginally significant at the $P < 0.1$ level (controversy and political inoculation conditions on irritation and uncertainty inoculation on distress, which were $P = 0.08$ for all three).

5.2 Domain specific policy support responses to politicization and inoculation

Prior research has shown decreases in support for relevant policy actions (such as state vaccine requirements) as a consequence of exposure to politicization.[20,36] Therefore, we expected that exposure to politicization would decrease support for HPV vaccine requirements in the case of the HPV vaccine and support for requiring masks in public places and businesses with respect to COVID-19. We find neither, as Figs. 3 and 4 show, respectively. Although all politicization treatments—across the three dimensions in the context of HPV and in combined form in the COVID-19 case—are negatively signed, there is no statistically significant difference between the base

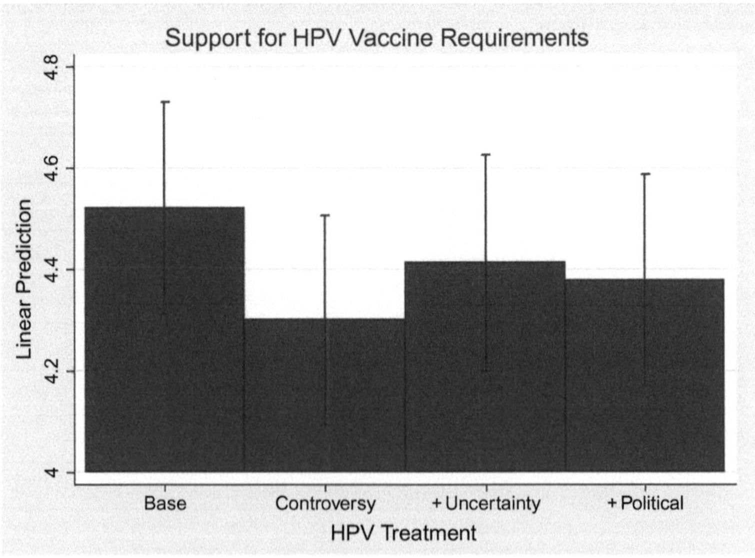

Fig. 3 Effects of politicization treatments on HPV vaccine requirements.

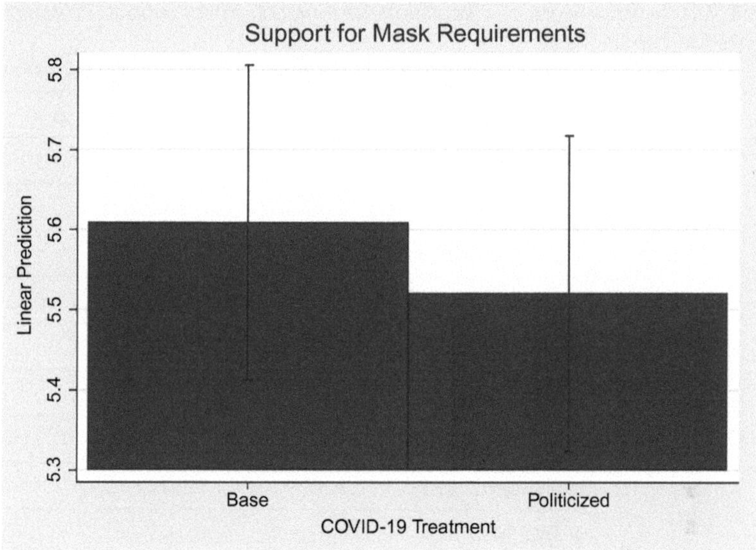

Fig. 4 Effects of politicization treatments on COVID-19 mask requirements.

level conditions and the politicized ones at standard levels of significance in a one-tailed test. Although the controversy condition is just outside the standard cut-off ($P=0.071$) in a comparison to the base condition, similar to the investigation with respect to emotions, we cannot statistically distinguish between the politicized dimensions in levels of policy support.

Although exposure to inoculation caused elevated negative emotional reactions among respondents, as described above, the pretreatment warnings do not seem to have significant effects on support for domain-specific policy action (see Table 4). Interestingly, across all four politicization comparisons, the inoculated conditions have higher mean support for policy action, which is the expected direction if inoculation were helping to blunt the negative effects of politicization; however, none are statistically significant by standard levels (the uncertainty condition P-value in the one-tailed test was 0.092, however).

5.3 Confusion and backlash responses to politicization and inoculation

Turning to confusion and backlash, we find strong evidence that exposure to any type of politicization increases confusion about the HPV vaccine (Fig. 5, left panel). Although the political condition has the highest mean, it is not statistically significantly different from either of the other two politicized

Table 4 *T*-test comparisons of inoculation warnings on policy support compared to politicization conditions alone.

		Support for domain-specific mandates		
		Mean	**SE**	**p**
HPV Vaccine	*Controversy*			0.40
	No Inoculation	4.30	0.104	
	+ Inoculation	4.34	0.106	
	+Uncertainty			0.092
	No Inoculation	4.41	0.106	
	+ Inoculation	4.60	0.098	
	+Political			0.21
	No Inoculation	4.38	0.109	
	+ Inoculation	4.5	0.106	
COVID-19	*Politicized*			0.19
	No Inoculation	5.61	0.100	
	+ Inoculation	5.73	0.088	

Note: One-tailed tests that inoculations are higher than regular treatment cells are reported.

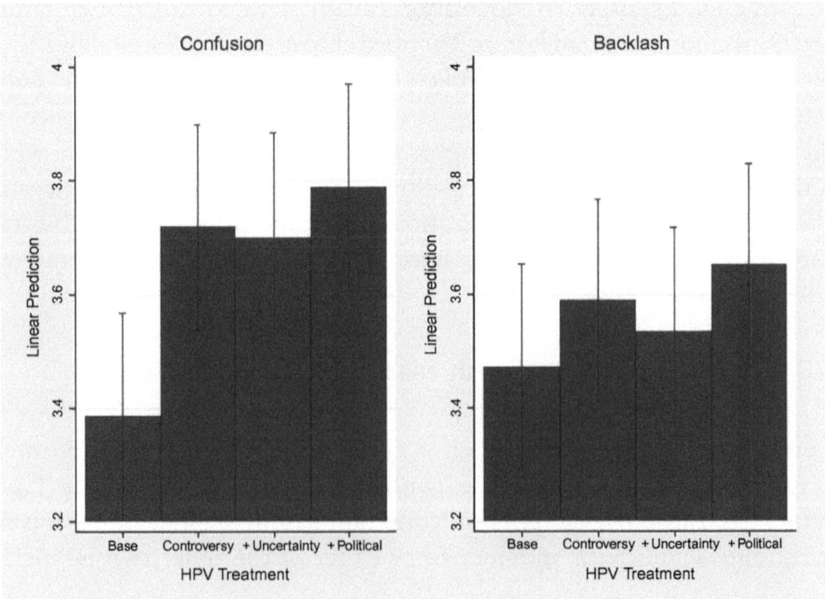

Fig. 5 Effects of politicization treatments on confusion and backlash (HPV vaccine).

conditions, suggesting that all three elicit similar responses. We do not find evidence that politicization causes backlash with respect to science at standard significance levels, although all three of the politicization conditions are positively signed—the expected direction if politicization were increasing backlash (Fig. 5, right panel)—and the political condition is marginally higher ($P=0.85$ in the one-tailed directional test compared to the base condition).

We find similar patterns as shown in Fig. 6 for both confusion and backlash (positive coefficients signaling an increase in both) in the context of the COVID-19 case, although the increase in backlash is just outside the standard cut-off for significance ($P=0.055$).

We do not find much support for the notion that inoculation could blunt the increased confusion, as none of the *t*-tests comparing means between the politicized treatments with and without inoculation are significant at standard levels. As shown in Table 5, we only find marginal support for a decrease in the political condition of the HPV vaccine at $P<0.1$ and in no other case. Therefore, we can mostly reject the hypothesis that inoculation can defend against increased confusion.

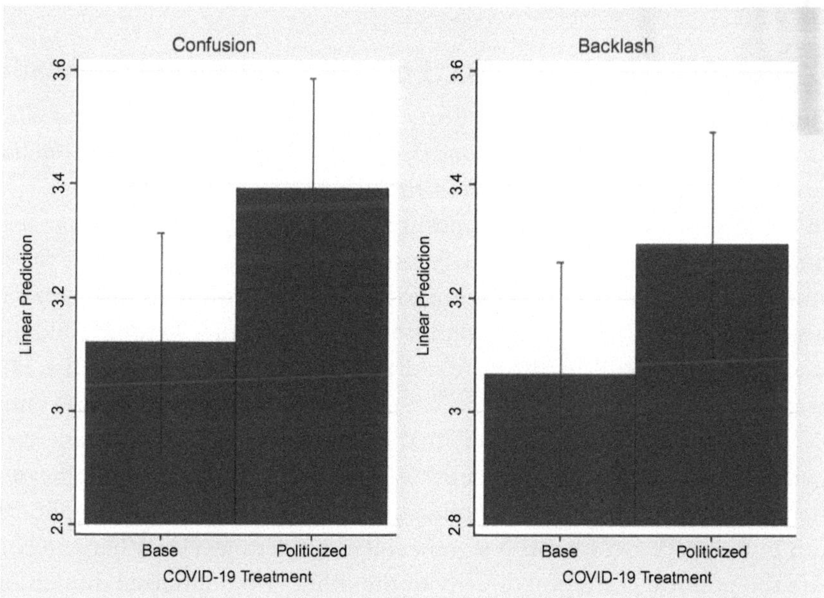

Fig. 6 Effects of politicization treatments on confusion and backlash (COVID-19).

Table 5 *T*-test comparisons of inoculation warnings on confusion and backlash compared to politicization conditions alone.

		Confusion			Backlash		
		Mean	SE	p	Mean	SE	p
HPV Vaccine	*Controversy*			0.66			0.81
	No Inoc.	3.72	0.088		3.59	0.087	
	+ Inoc.	3.77	0.090		3.70	0.091	
	+Uncertainty			0.30			0.25
	No Inoc.	3.70	0.094		3.53	0.089	
	+ Inoc.	3.63	0.93		3.45	0.92	
	+Political			0.095			0.18
	No Inoc.	3.79	0.096		3.65	0.097	
	+ Inoc.	3.61	0.092		3.53	0.094	
COVID-19	*Politicized*			0.18			0.44
	No Inoc.	3.92	0.096		3.29	0.102	
	+ Inoc.	3.26	0.108		3.27	0.108	

Note: One-tailed tests that inoculations are lower than regular treatment cells are reported.

5.4 Potential carryover effects of politicization and inoculation

Turning to potential carryover effects, we first examine the extent to which confusion over politicization in one domain (HPV vaccine or masks) spills or carries over to influence confusion more broadly. As shown in Fig. 7, we find that the coefficients for exposure to all types of politicization (each dimension with respect to the HPV vaccine treatments as shown in the left panel and the combined dimension for COVID-19 on the right panel) are positive, indicating increased confusion over "whether getting childhood vaccinations is good or bad for children's health" in the case of the HPV treatment and "whether there are effective strategies for preventing COVID-19 (coronavirus) spread" in the case of the COVID-19 masks treatment. While the increase in confusion carryover is statistically significant in the COVID-19 case, only the political cues operationalization of politicization in the HPV treatments is significant at standard levels. When we compare the political condition directly to the other two politicized dimensions,

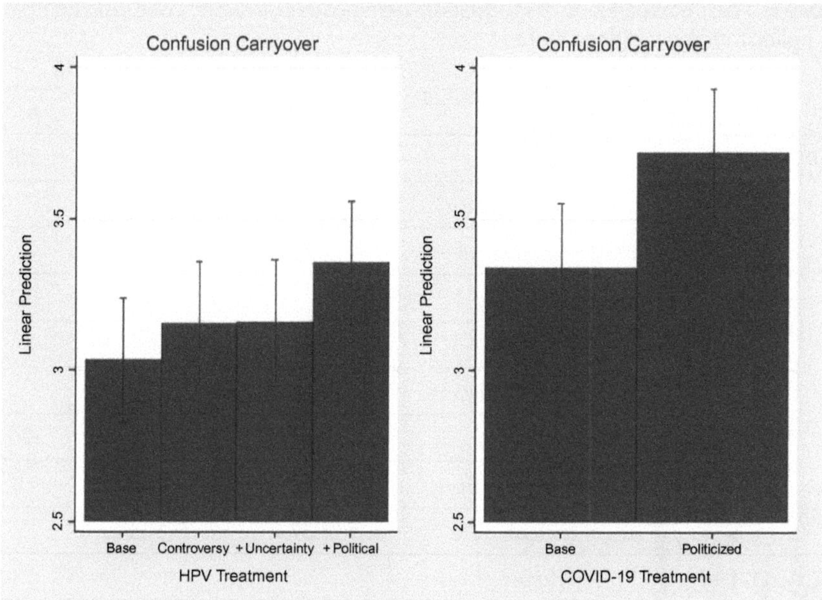

Fig. 7 Effects of politicization treatments on carryover confusion (HPV vaccine and COVID-19).

however, the levels are not statistically different from each other, even at the $P<0.1$ threshold in a two-tailed test. Turning to the COVID-19 context, we see that the politicized condition is statistically higher than the base case in confusion carryover.

In comparing the conditions with and without the inoculation messaging for both the political dimension for HPV and all elements of politicization for COVID-19 alone, we find some evidence that inoculation may blunt the increase in confusion (Table 6). With inoculation, the HPV political condition mean is 3.1 compared to 3.4 without it, which is statistically significant in a one-tailed test for a decrease ($P=0.03$). These results suggest that although the political condition may increase confusion, it does appear to be the case that inoculation against political controversy can mitigate the effects. In the case of COVID-19, the mean for the politicization treatment with inoculation is also lower than the mean without inoculation (3.5 compared to 3.7) although the one-tailed test is outside of traditional standards for significance ($P=0.08$).

To examine another potential marker of carryover effects, we ask whether reading about politicization decreases support for immunization programs more broadly. As shown in Fig. 8, exposure to politicization

Table 6 *T*-test comparisons of inoculation warnings on carryover confusion compared to politicization conditions alone.

		Carryover confusion		
		Mean	**SE**	**p**
HPV Vaccine	*Controversy*			0.87
	No Inoculation	3.16	0.095	
	+ Inoculation	3.31	0.103	
	+Uncertainty			0.36
	Treatment	3.16	0.105	
	with Inoculation	3.11	0.101	
	+Political			0.029
	No Inoculation	3.36	0.109	
	+ Inoculation	3.07	0.104	
COVID-19	*Politicized*			0.077
	No Inoculation	3.72	0.108	
	+ Inoculation	3.50	0.112	

Note: One-tailed tests that inoculations are lower than regular treatment cells are reported.

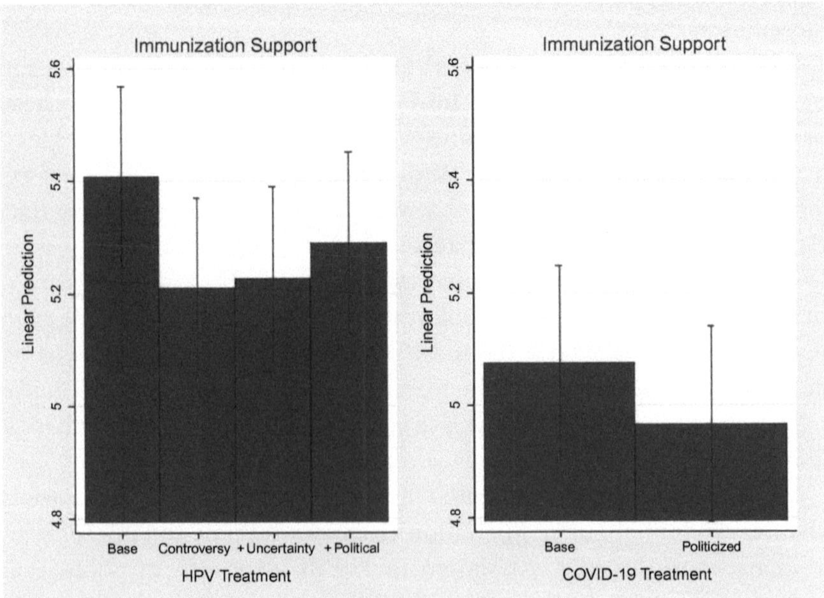

Fig. 8 Effects of politicization treatments on carryover support for immunization program requirements (HPV vaccine and COVID-19).

Table 7 T-test comparisons of inoculation warnings on carryover immunization support compared to politicization conditions alone.

		Immunization support		
		Mean	**SE**	**p**
HPV Vaccine	*Controversy*			0.25
	No Inoculation	5.21	0.079	
	+ Inoculation	5.29	0.077	
	+Uncertainty			0.23
	No Inoculation	5.23	0.086	
	+ Inoculation	5.31	0.079	
	+Political			0.30
	No Inoculation	5.29	0.084	
	+ Inoculation	5.35	0.082	
COVID-19	*Politicized*			0.11
	No Inoculation	4.97	0.091	
	+ Inoculation	5.12	0.089	

Note: One-tailed tests that inoculations are higher than regular treatment cells are reported.

(whether one of the dimensions in the HPV vaccine context or to the all version in the COVID-19 context) consistently results in negative coefficients, suggesting that it lowers support for immunization programs. However, only the HPV vaccine politicization condition operationalized as controversy alone is significant at standard levels in a one-tailed test for a decrease ($P=0.04$). (Similar to earlier analyses, however, this condition is not statistically distinguishable from the uncertainty or political dimension results).

Turning to the effect of inoculation messaging, none of the inoculation comparisons are statistically different from their counterparts without inoculation, even though all conditions with inoculation have a mean of immunization program support that is greater than the conditions without inoculation, the expected direction for protective effects (Table 7).

6. Discussion

In this analysis, we find that exposure to media messages emphasizing politicization increases negative emotional reactions with respect to irritation and distress in particular. Reading about COVID-19 generally was particularly likely to increase negative emotional arousal; reading politicization messages on this high salience issue was even more likely to exacerbate these negative emotions. Although our study does not identify evidence of domain-specific decreases in policy support as a result of politicization messages, we do find support for the notion that politicization increases confusion both on the topical domain in question and in potential carryover domains (that is, confusion about broader health recommendations). For the latter, our findings suggest that politicization could have a deleterious influence on broader health-related attitudes. Although our results are consistent in sign with prior work finding that exposure to conflicting evidence about health produces heightened backlash,[16] we do not find support for substantial increases in backlash as a result of any type of politicization in this study. Similarly, although results are in the expected direction, we find little support for potential carryover effects on broader policy attitudes (support for general immunization programs, in this case), which is a reassuring finding although certainly not definitive evidence on the matter.

One goal of this large pilot study was to assess whether, through an experimental design, the various conceptual ways scholars have operationalized politicization have any differential effects. Previous observational studies have shown that the public does distinguish among controversy, evidence uncertainty, and political discourse or cues, and that perceptions of evidence uncertainty are correlated with reduced policy support.[19,20] However, experimental evidence is needed to evaluate the causal effects of these dimensions. By and large this pilot study does not suggest that any one operationalization of politicization is more or less damaging than the others. Rather, the findings suggest that all types of politicization are equally concerning, particularly as judged by their effects on negative emotions and confusion. There are some instances in which the political operationalization (the reference to a politician) elicits the strongest reaction compared to the base information, albeit not statistically different from the other dimensions, and thus the presence of political officials in health media coverage is worthy of future exploration. Given that politicization of the COVID-19 pandemic was well underway at the time of fielding,[7] it seems especially important to acknowledge that we still uncovered effects of additional exposure to the experimental politicization message, regardless of the extent to which some respondents may have been pretreated with exposure to politicization before taking our survey.[18,50]

One important innovation of this study relative to prior work is that we tested a potential message strategy to mitigate any adverse effects of exposure to media messages about politicization that was tailored to the particular dimension of politicization. Can providing such a tailored warning prior to exposure help to reduce some of its negative effects? We find some evidence that inoculation can help blunt the adverse effects of exposure to politicization, especially with respect to carryover confusion. However, it is notable that we largely do not find that inoculation can help mitigate confusion on the topic-specific domain. In addition, it is important to acknowledge that respondents became more agitated when we provided a warning in advance of exposure to politicization, suggesting that this particular inoculation strategy is not a costless intervention, in that it heightened irritation and distress among participants.

Our analysis is not without limitations. First and foremost, our sample—although recruited nationally (rather than in just one geographic area) and diverse in demographic characteristics like age, gender and race—is not nationally representative. The sample (see Table A1 in Appendix) is somewhat more educated than average, and because the survey was fielded exclusively online, respondents all had Internet access. These characteristics may affect our ability to generalize the findings to the broader population. However, because we used a randomized experimental design, the sample composition does not affect our ability to detect causal effects of our politicization and inoculation treatments. Second, although we expect that several factors would moderate the effects of politicization (including education, attitudes toward science, and partisanship/ideology, to name a few), analyzing potential subgroup effects is beyond the scope of the current pilot investigation. We did not power this pilot study to ensure sufficient numbers within subgroups; we conceptualized this mainly as a pilot of the main effects of politicization dimension treatments and inoculation. Further, as noted above, given we fielded the experiment during the summer of 2020 in the heat of the early days of the COVID-19 pandemic, pretreatment exposure to politicized controversy, amplification of scientific uncertainty, and partisan intrusion in science was likely a broad influence on respondents. It also could have been unequally distributed among respondents in ways that could have affected our experiment (i.e., distributed according to media exposure, or by trust in science) since we cannot easily test for randomization conditioned on prior exposure. If anything, however, we think these factors may have made it harder to detect effects of an additional treatment in our study. Finally, our political cues treatments only included a reference to "politicians," and we did not provide a specific partisan cue. However, we would anticipate that inclusion of partisan cues may be particularly potent influences on respondents based on past research.[52,53] (Although, evidence is uncertain on the power of partisan cues on COVID-19 attitudes in particular, perhaps because the partisan characteristics of COVID-19 were already so well understood in public discourse[54]). Regardless, future experimental tests of the various dimensions of politicization should consider partisan cueing in particular, to test for uniform or asymmetric effects of partisan rhetoric.

Overall, our results suggest that while different types of messages might convey a politicized context to the public in media coverage, our findings as they relate to the HPV vaccine context do not suggest that one type of politicization is necessarily more pernicious than the others. In fact, all types of politicization appear to foster negative responses among the public, particularly arousing negative emotions and confusion, on both the topics in question and on broader health recommendations. Finally, our findings offer suggestive evidence that one type of communication strategy, a warning about actors' strategic use of politicization, can sometimes and partially blunt the adverse effects of exposure to different types of politicization, although not without concerns. Future research on more types of strategies (including interventions in media, education, and other social domains) are needed given the increasing frequency of politically-charged and controversial messaging in health and science discourse.

Appendix

A.1 Appendix: Demographic sample

Table A1 Demographics of the survey experiment sample

		Number of respondents	Percentage
Gender	Female	1616	53.7%
Age	18–24	355	11.8%
	25–34	465	15.5%
	35–44	547	18.2%
	45–54	522	17.3%
	55–64	468	15.6%
	65+	653	21.7%
Race	Nonwhite	1073	35.6%
Education	HS or less	597	19.8%
	Some college	844	28.0%
	College and above	1571	52.2%

A.2 Appendix: Treatment language

A.2.1 HPV conditions

{Headline for condition:}

{Base:}	{Controversy:}	{Uncertainty:}	{Political:}
Panel to Expand HPV Vaccine Recommendations	Controversy Erupts Over New HPV Vaccine Recommendations	Advocates Cast Doubt on Scientific Evidence Behind New HPV Vaccine Recommendations as Controversy Erupts	Political Pushback to New HPV Vaccine Recommendations as Controversy Erupts

August 2019—The Advisory Committee on Immunization Practices (ACIP) updated its human papillomavirus (HPV) vaccine guidelines in June. The committee now recommends that both men and women up to age 26 get a "catch-up" vaccine if they missed the shots in preadolescence, when it is usually offered. The vaccine protects against HPV, a virus that is commonly spread through sex and can cause cervical and other cancers and genital warts. ACIP panelist and adolescent medicine specialist Dr. Anna Rogers said, "These guidelines are based on scientific evidence that the vaccine is extremely effective at preventing many types of cancers. In fact, a new study found a significant decrease in the prevalence of two strains of HPV that cause most cervical cancer in countries that recommend HPV vaccination, which has exceeded expectations."

The vaccine is approved for people up to age 45, but the same panel declined a proposal to recommend it for people older than 26, settling on a weak endorsement for adults between 26 and 45—which means patients and doctors can make the decision together. The CDC recommends that children receive the first dose of the HPV vaccine between 11 and 12. However, vaccination rates for HPV in the United States are low in part because most states do not mandate the HPV vaccine as a requirement for school entry, even though many states have considered it.

{Politicization treatments add the following to all conditions:}

Heated debate erupted soon after the controversial new ACIP recommendations were released. Proponents of broader vaccination argued that more states should consider requiring the vaccine for youth, given evidence of the vaccine's effectiveness in preventing cancer. Opponents of required vaccination sharply disagree and are using the new recommendations to argue against more state requirements, since they believe that the vaccines expose youth to side effects unnecessarily.

{Then politicization treatments add the following as specified by the condition:}.

{Controversy:}	{Uncertainty:}	{Political:}
Local advocate Jennifer Peters summed up the concerns: "Why are we requiring 10- and 11-year old kids to get a vaccine that they can get later as adults?"	In addition, local advocate Jennifer Peters cast doubt on the evidence provided for the recommended age to receive vaccination. "Researchers are always changing their minds about when to vaccinate, and the evidence is uncertain and can be used to support different positions. In this case, they know that the vaccine is fine for adults and yet they say that it has to be given to children to be effective." Peters summed up the concerns: "Why are we requiring 10- and 11-year old kids to get a vaccine that they can get later as adults?"	Longtime politician and state senator Jennifer Peters sided with opponents of required vaccination, summing up the concerns: "Why are we requiring 10- and 11-year old kids to get a vaccine that they can get later as adults?"

A.2.2 COVID-19 conditions

{Headline for condition:}

{Base:}	{All Politicized Elements:}
CDC Updates Recommendations on Mask Wearing	Controversy and Political Pushback Over CDC Mask Wearing Recommendations in Light of Doubts About Scientific Evidence

June 2020—The Centers for Disease Control and Prevention updated its COVID-19 (coronavirus) recommendations in June, reiterating guidelines on mask wearing that first came out in April. The CDC and the World

Health Organization now recommend that people wear cloth face coverings when in public as an additional, voluntary public health measure. The mask protects the community against the spread of coronavirus infection by reducing the extent to which an infected individual will contaminate the air and surfaces in public areas. CDC official and infectious disease specialist Dr. Anna Rogers said, "These guidelines are based on scientific evidence that mask wearing in public is an effective way of preventing respiratory droplet spread. In fact, studies show people without symptoms and people who later develop symptoms can spread the virus unknowingly, which was a driving factor in the updating the guidelines."

The CDC also noted reports from other countries showing that a cloth face covering can reduce the chance of spreading the virus in public settings where social distancing is difficult. Social distancing refers to keeping space between you and others and limiting contact with others outside your household in indoor and outdoor spaces. As state economies open back up, these preventative measures may become more important.

{Politicized treatment adds the following:}

Heated debate erupted soon after the controversial CDC recommendations were released. Proponents of broader mask usage argued that states should consider requiring masks, given evidence of their effectiveness in preventing droplet spread. Opponents of mask usage sharply disagree, pointing to recent comments by a World Health Organization spokesperson, who stated that much is unclear about how much asymptomatic spread there actually is by people who don't know they are infected.

In particular, longtime politician and state senator Jennifer Peters cast doubt on the scientific evidence that the CDC used to support changing the guidelines. "Researchers are always changing their minds about how to prevent coronavirus spread, and the evidence is uncertain and can be used to support different positions." Citing partisan disagreement between Democrats and Republicans and President Trump's public declaration that he won't wear a mask, Peters summed up the concerns: "Why are we infringing on people's rights when there is little actual agreement that mask wearing actually helps?"

References

1. Suhay E, Druckman JN. The politics of science: Political values and the production, communication, and reception of scientific knowledge. *Ann Am Acad Pol Soc Sci*. 2015;658(1):6–15. https://doi.org/10.1177/0002716214559004.
2. Hart PS, Nisbet EC. Boomerang effects in science communication: How motivated reasoning and identity cues amplify opinion polarization about climate mitigation policies. *Commun Res*. 2012;39(6):701–723.

3. Gauchat G. Politicization of science in the public sphere: A study of public Trust in the United States, 1974 to 2010. *Am Sociol Rev.* 2012;77(2):167–187. https://doi.org/10.1177/0003122412438225.
4. Lupia A. What is the value of social science? Challenges for researchers and government funders. *PS Polit Sci Amp Polit.* 2014;47(1):1–7. https://doi.org/10.1017/S1049096513001613.
5. Motta M. The dynamics and political implications of anti-intellectualism in the United States. *Am Politics Res.* 2018;46(3):465–498. https://doi.org/10.1177/1532673X17719507.
6. Bolsen T, Druckman JN. Counteracting the politicization of science. *J Commun.* 2015;65(5):745–769. https://doi.org/10.1111/jcom.12171.
7. Gadarian SK, Goodman SW, Pepinsky TB. Partisanship, health behavior, and policy attitudes in the early stages of the COVID-19 pandemic. *PLoS One.* 2021;16(4):e0249596. https://doi.org/10.1371/journal.pone.0249596.
8. Hart PS, Chinn S, Soroka S. Politicization and polarization in COVID-19 news coverage. *Sci Commun.* 2020. https://doi.org/10.1177/1075547020950735. Published online August 25.
9. Muddiman A, Buden C, Romas B, et al. Cable and nightly network news coverage of coronavirus. *Center for Media Engagement.* 2020. https://mediaengagement.org/research/coronavirus-network-coverage.
10. May T. Anti-Vaxxers, politicization of science, and the need for Trust in Pandemic Response. *J Health Commun.* 2020;25(10):761–763. https://doi.org/10.1080/10810730.2020.1864519.
11. Bokemper SE, Huber GA, Gerber AS, James EK, Omer SB. Timing of COVID-19 vaccine approval and endorsement by public figures. *Vaccine.* 2021;39(5):825–829. https://doi.org/10.1016/j.vaccine.2020.12.048.
12. Kreps SE, Kriner DL. Model uncertainty, political contestation, and public trust in science: Evidence from the COVID-19 pandemic. *Sci Adv.* 2020;6(43): eabd4563. https://doi.org/10.1126/sciadv.abd4563.
13. Allcott H, Boxell L, Conway J, Gentzkow M, Thaler M, Yang D. Polarization and public health: Partisan differences in social distancing during the coronavirus pandemic. *J Public Econ.* 2020;191. https://doi.org/10.1016/j.jpubeco.2020.104254, 104254.
14. Gollust SE, Nagler RH, Fowler EF. The emergence of COVID-19 in the U.S.: A public health and political communication crisis. *J Health Polit Policy Law.* 2020. https://doi.org/10.1215/03616878-8641506.
15. Bavel JJV, Baicker K, Boggio PS, et al. Using social and behavioural science to support COVID-19 pandemic response. *Nat Hum Behav.* 2020;4(5):460–471. https://doi.org/10.1038/s41562-020-0884-z.
16. Nagler RH, Vogel RI, Gollust SE, Yzer MC, Rothman AJ. Effects of prior exposure to conflicting health information on responses to subsequent unrelated health messages: Results from a population-based longitudinal experiment. *Ann Behav Med.* 2021, kaab069. https://doi.org/10.1093/abm/kaab069.
17. Lupia A. *Uninformed: Why People Seem to Know So Little about Politics and What we Can Do about it.* Oxford University Press; 2015.
18. Fowler EF, Gollust SE. The content and effect of politicized health controversies. *Ann Am Acad Pol Soc Sci.* 2015;658(1):155–171. https://doi.org/10.1177/0002716214555505.
19. Fowler EF, Nagler RH, Gollust SE. Public views on the politicization of health and science: How the public assesses controversy, scientific uncertainty and political discourse for nine issues. In: *Annual Meeting of the Midwest Political Science Association*; 2017.
20. Saulsberry L, Fowler EF, Nagler RH, Gollust SE. Perceptions of politicization and HPV vaccine policy support. *Vaccine.* 2019;37(35):5121–5128.

21. Prusaczyk B, Swindle T, Curran G. Defining and conceptualizing outcomes for de-implementation: Key distinctions from implementation outcomes. *Implement Sci Commun*. 2020;1(1):43. https://doi.org/10.1186/s43058-020-00035-3.
22. Ubel PA, Asch DA. Creating value in health by understanding and overcoming resistance to De-innovation. *Health Aff (Millwood)*. 2015;34(2):239–244. https://doi.org/10.1377/hlthaff.2014.0983.
23. Patashnik EM, Gerber AS, Dowling CM. *Unhealthy Politics*. Princeton University Press; 2017. https://press.princeton.edu/books/hardcover/9780691158815/unhealthy-politics. Accessed 11 November 2021.
24. Lantz PM, Evans WD, Mead H, Alvarez C, Stewart L. Knowledge of and attitudes toward evidence-based guidelines for and against clinical preventive services: Results from a National Survey. *Milbank Q*. 2016;94(1):51–76. https://doi.org/10.1111/1468-0009.12181.
25. Carman KL, Maurer M, Yegian JM, et al. Evidence that consumers are skeptical about evidence-based health care. *Health Aff (Millwood)*. 2010;29(7):1400–1406. https://doi.org/10.1377/hlthaff.2009.0296.
26. Evans WD, Lantz PM, Mead K, Alvarez C, Snider J. Adherence to clinical preventive services guidelines: Population-based online randomized trial. *SSM - Popul Health*. 2015;1:48–55. https://doi.org/10.1016/j.ssmph.2015.11.003.
27. Schlesinger M, Grob R. Treating, fast and slow: Americans' understanding of and responses to low-value care. *Milbank Q*. 2017;95(1):70–116. https://doi.org/10.1111/1468-0009.12246.
28. Nagler RH, Fowler EF, Gollust SE. Covering controversy: What are the implications for Women's health? *Womens Health Issues*. 2015;25(4):318–321. https://doi.org/10.1016/j.whi.2015.04.011.
29. Gauchat G. The political context of science in the United States: Public acceptance of evidence-based policy and science funding. *Soc Forces*. 2015;94(2):723–746. https://doi.org/10.1093/sf/sov040.
30. Bolsen T, Palm R, Kingsland JT. Counteracting climate science politicization with effective frames and imagery. *Forensic Sci Commun*. 2019;41(2):147–171. https://doi.org/10.1177/1075547019834565.
31. Lewandowsky S, Oberauer K. Motivated rejection of science. *Curr Dir Psychol Sci*. 2016;25(4):217–222. https://doi.org/10.1177/0963721416654436.
32. Druckman JN, Klar S, Krupnikov Y, Levendusky M, Ryan JB. How affective polarization shapes Americans' political beliefs: A study of response to the COVID-19 pandemic. *J Exp Polit Sci*. 2020;1–12. Published online. https://doi.org/10.1017/XPS.2020.28.
33. Green J, Edgerton J, Naftel D, Shoub K, Cranmer SJ. Elusive consensus: Polarization in elite communication on the COVID-19 pandemic. *Sci Adv*. 2020;6(28). https://doi.org/10.1126/sciadv.abc2717, eabc2717.
34. Dunwoody S, Kohl PA. Using weight-of-experts messaging to communicate accurately about contested science. *Sci Commun*. 2017;39(3):338–357. https://doi.org/10.11771075547017707765.
35. Bolsen T, Druckman JN, Cook FL. How frames can undermine support for scientific adaptations: Politicization and the status-quo bias. *Public Opin Q*. 2014;78(1):1–26. https://doi.org/10.1093/poq/nft044.
36. Gollust SE, Dempsey AF, Lantz PM, Ubel PA, Fowler EF. Controversy undermines support for state mandates on the human papillomavirus vaccine. *Health Aff (Millwood)*. 2010;29(11):2041–2046. https://doi.org/10.1377/hlthaff.2010.0174.
37. Lerman AE, Sadin ML, Trachtman S. Policy uptake as political behavior: Evidence from the affordable care act. *Am Polit Sci Rev*. 2017;111(4):755–770. https://doi.org/10.1017/S0003055417000272.

38. Trachtman S. Polarization, participation, and premiums: How political behavior helps explain where the ACA works, and where it Doesn't. *J Health Polit Policy Law.* 2019;44(6):855–884. https://doi.org/10.1215/03616878-7785787.
39. Sances MW, Clinton JD. Who participated in the ACA? Gains in insurance coverage by political partisanship. *J Health Polit Policy Law.* 2019;44(3):349–379. https://doi.org/10.1215/03616878-7366988.
40. Nagler RH, Yzer MC, Rothman AJ. Effects of media exposure to conflicting information about mammography: Results from a population-based survey experiment. *Ann Behav Med.* 2019;53(10):896–908. https://doi.org/10.1093/abm/kay098.
41. Chang C. Motivated processing: How people perceive news covering novel or contradictory Health Research findings. *Sci Commun.* 2015;37(5):602–634. https://doi.org/10.1177/1075547015597914.
42. Lyons BA, Merola V, Reifler J. Shifting medical guidelines: Compliance and spillover effects for revised antibiotic recommendations. *Soc Sci Med.* 2020;255: 112943. https://doi.org/10.1016/j.socscimed.2020.112943.
43. Lupia A. Communicating science in politicized environments. *Proc Natl Acad Sci.* 2013;110(Supplement 3):14048–14054. https://doi.org/10.1073/pnas.1212726110.
44. Han J, Kim Y. Defeating merchants of doubt: Subjective certainty and self-affirmation ameliorate attitude polarization via partisan motivated reasoning. *Public Underst Sci.* 2020;29(7):729–744. https://doi.org/10.1177/0963662520939315.
45. Niederdeppe J, Gollust SE, Barry CL. Inoculation in competitive framing: Examining message effects on policy preferences. *Public Opin Q.* 2014;78(3):634–655.
46. Eagly AH, Chaiken S. *The Psychology of Attitudes.* Harcourt Brace Jovanovich College Publishers; 1993.
47. Nyhan B, Reifler J. When corrections fail: The persistence of political misperceptions. *Polit Behav.* 2010;32(2):303–330. https://doi.org/10.1007/s11109-010-9112-2.
48. McGuire WJ. Inducing resistance to persuasion. Some contemporary approaches. In: Haaland CC, Kaelber WO, eds. *Self and Society. An Anthology of Readings, Lexington Mass.* Ginn Cust Publishing; 1981:192–230. Published online 1964. Accessed October 4, 2021. https://opus4.kobv.de/opus4-Fromm/frontdoor/index/index/docId/16094.
49. Compton JA, Pfau M. Inoculation theory of resistance to influence at maturity: Recent Progress in theory development and application and suggestions for future research. *Ann Int Commun Assoc.* 2005;29(1):97–146. https://doi.org/10.1080/23808985.2005.11679045.
50. Druckman JN, Leeper TJ. Learning more from political communication experiments: Pretreatment and its effects. *Am J Polit Sci.* 2012;56(4):875–896.
51. Gollust SE, Fowler EF, Nagler RH. Prevalence and potential consequences of exposure to conflicting information about mammography: Results from nationally-representative survey of U.S. *Adults Health Commun.* 2021;1–14. https://doi.org/10.1080/10410236.2021.1951958.
52. Bisgaard M, Slothuus R. Partisan elites as culprits? How party cues shape partisan perceptual gaps. *Am J Polit Sci.* 2018;62(2):456–469. https://doi.org/10.1111/ajps.12349.
53. Jerit J, Barabas J. Partisan perceptual bias and the information environment. *J Theor Polit.* 2012;74(3):672–684.
54. Gadarian SK, Goodman SW, Pepinsky T. Partisan endorsement experiments do not affect mass opinion on COVID-19. *J Elections Public Opin Parties.* 2021;31(sup1):122–131. https://doi.org/10.1080/17457289.2021.1924727.

CHAPTER SEVEN

Vax attacks: How conspiracy theory belief undermines vaccine support

Christina E. Farhart[a,*], Ella Douglas-Durham[b], Krissy Lunz Trujillo[c,d], and Joseph A. Vitriol[e]

[a]Carleton College, Department of Political Science, Northfield, MN, United States
[b]Harvard T.H. Chan School of Public Health, Department of Social and Behavioral Sciences, Cambridge, MA, United States
[c]Harvard Kennedy School, Shorenstein Center, Cambridge, MA, United States
[d]Northeastern University, Network Science Institute, Boston, MA, United States
[e]Stony Brook University, Department of Political Science, Stony Brook, NY, United States
*Corresponding author: e-mail address: cfarhart@carleton.edu

Contents

1. Introduction 136
2. Literature review and framework 137
   2.1 General vaccine hesitancy and uptake 137
   2.2 Influence of misinformation and conspiracy theories on vaccine hesitancy 139
   2.3 Expected correlates of COVID-19 vaccine hesitancy 141
3. Hypotheses 143
4. Data and methods 143
   4.1 Data 143
   4.2 Measures 144
   4.3 Dependent variable 144
   4.4 Independent variables 144
5. Results 148
   5.1 Effect of conspiracy belief on vaccine support and hesitancy 148
   5.2 Effect of political identity on vaccine support and hesitancy 153
   5.3 Effect of anti-intellectualism on vaccine support and hesitancy 153
   5.4 Comparative effects 153
6. Discussion and conclusion 155
Appendix A. Question Wording 157
   A.1 Vaccine hesitancy 158
   A.2 COVID-19 conspiracy theory belief 158
   A.3 Conspiracy predisposition/conspiratorial thinking 160
   A.4 Partisanship 160
   A.5 Political ideology 160
   A.6 Anti-intellectualism 161
   A.7 Trust 161

Progress in Molecular Biology and Translational Science, Volume 188
ISSN 1877-1173
https://doi.org/10.1016/bs.pmbts.2021.11.001

A.8 Internal efficacy 161
A.9 External efficacy 161
A.10 Learned helplessness 162
A.11 Satisfaction 162
A.12 Political knowledge 162
A.13 Political interest 163
A.14 Age 163
A.15 Race 163
A.16 Gender 164
A.17 Education 164
A.18 Income 164
References 164

Abstract

As the world continues to respond to the spread of a novel coronavirus (SARS-CoV-2, which causes the disease commonly known as COVID-19), it has become clear that one of the most effective strategies for curbing the pandemic is the COVID-19 vaccine. However, a major challenge that health organizations face when advocating for the uptake of the COVID-19 vaccine is the spread of related misinformation and conspiracy theories. This study examines factors that influence vaccine hesitancy using two online survey samples, one convenience and one nationally representative, collected in the early summer of 2020 during the height of the second peak of coronavirus cases in the United States. Given extant literature on vaccine hesitancy and conspiracy belief, we expect that three factors—conspiracy theory belief, political identity, and anti-intellectualism—have served to reduce COVID-19 vaccination likelihood. Accordingly, across our two independent samples we find that anti-intellectualism, conspiratorial predispositions, and COVID-19 conspiracy theory belief are the strongest and most consistent predictors of COVID-19 vaccine hesitancy. Notably, we also find that partisanship and political ideology are inconsistently significant predictors of COVID-19 vaccine hesitancy once conspiracy theory beliefs, anti-intellectualism, and control variables are accounted for in the models. When political tendencies are significant, they demonstrate a relatively small substantive association with COVID-19 vaccine hesitancy. We discuss implications for ongoing mass vaccination efforts, continued widespread vaccine hesitancy, and related political attitudes.

1. Introduction

Vaccine hesitancy has existed since vaccines were first developed[1] and represents an ongoing threat to global public health. It can decrease vaccination rates both directly through lessening vaccine uptake and indirectly through hostility to mandatory vaccination policies.[2,3] Decreases in vaccination rates have contributed to the increased prevalence of preventable deaths and the reappearance of previously eradicated diseases such as polio and measles.[2] Vaccine hesitancy challenges public health to such an extent that the World Health Organization (WHO) called it 1 of the 10 biggest threats to global health.[4]

In light of the COVID-19 pandemic, vaccine hesitancy is a particularly salient concern. In the United States, approximately a third of Americans do not plan on getting the COVID-19 vaccine.[5] In other countries, COVID-19 vaccine acceptance rates vary widely.[6] Such widespread vaccine hesitance jeopardizes the pandemic recovery through insufficiently high vaccination rates, which in turn can contribute to the rise of vaccine-resistant strains.[7]

Given the seriousness of the issue, here we focus on understanding the role of three predicted contributors to vaccine hesitancy: partisanship, anti-intellectualism, and conspiracy theory predisposition and belief. Using two original online samples of U.S. adults collected through MTurk and Forthright, the latter of which was sampled to approximate national representativeness, we find that anti-intellectualism, conspiratorial predispositions, and COVID-19 conspiracy theory belief are the strongest and most consistent predictors of vaccine hesitancy. Notably, we also find that while partisanship and ideological self-placement are statistically significant predictors of vaccine hesitancy, once conspiracy theory, anti-intellectualism, and control variables are accounted for in the models, these relationships are no longer significant. When political tendencies are significant in these full models, they hold a relatively small substantive association with vaccine hesitancy.

In what follows, we first discuss what vaccine hesitancy is and what tends to drive vaccine hesitancy generally. Then, we discuss the role of misinformation, conspiracy theories, partisanship, and anti-intellectualism in vaccine hesitancy. Next, we apply this discussion to expectations surrounding COVID-19. Finally, we report direct tests of our hypotheses, and then discuss our results and their implications for on-going mass vaccination efforts, continued widespread vaccine hesitancy, and related political attitudes.

2. Literature review and framework

2.1 General vaccine hesitancy and uptake

In this section, we review contributing factors to vaccine hesitancy. In doing so we provide context for the role of misinformation and conspiracy theory spread as drivers of increased vaccine hesitancy. Vaccine hesitancy can be defined as "a delay in acceptance or refusal of vaccination despite availability of vaccine services."[8] Others have extended this definition to include people who accept or receive some vaccines but remain concerned about their safety, effectiveness, or importance.[9] Here, we use vaccine support as the inverse of vaccine hesitancy. Importantly, vaccine hesitancy is multifaceted and can vary across context, population, and vaccine type, although a more generalized predisposition toward hesitancy has been observed.[8,10] Nonetheless, factors associated with increased vaccine hesitancy can be classified into three broad categories: convenience, complacency, and confidence.[8]

The convenience of getting vaccines is one factor determining vaccine hesitancy levels.[8] For instance, it may be difficult to access vaccines in rural areas or places far from hospitals, clinics, or other providers.[11] Further, it may be inconvenient to get inoculated if it is financially costly; in this case, systems that have expensive health insurance, costly medical care, or other similar issues can increase vaccine hesitancy, particularly for the poor and disadvantaged.[12] Complacency and confidence are reflected in concerns over vaccine safety, effectiveness, and importance.[13,14] Complacency encompasses concerns over importance, i.e., that people do not feel urgency or perceive risk in the vaccine-preventable disease.[15,16] For example, some Americans underestimated the threat of COVID-19 during the pandemic, prompting reduced compliance with CDC guidelines for reducing the risk of contagion.[17] Other research indicates that as Americans have lived through the pandemic, increased familiarity of the disease has coincided with decreased perceptions of threat.[18] This desensitization to the pandemic, alongside misinformation about COVID-19, have led to increased vaccine hesitancy over time.[18] Finally, vaccine confidence centers around concerns about safety and, to a lesser extent, effectiveness. Often individuals are worried about potential side effects, harmful "ingredients," inefficacy, *and* that such information about these potential concerns are being withheld or unacknowledged.[10,19,20] Such fears are compounded when individuals do not trust various experts and institutions, including doctors, scientists, and the pharmaceutical industry.[14,21]

Vaccine hesitancy or anti-vaccine stances covary with several individual-level and group-based factors. For example, higher levels of vaccine hesitancy are observed among people who experience particular predispositions such as, higher conspiratorial thinking, blood and needle sensitivity, individualism, disgust sensitivity,[22,23,24] injection fear,[25] and moral purity.[26] A study across 25 national samples in 12 countries found that being female, having trust in medical and scientific experts, and worry about COVID-19 are the most consistent and robust predictors of willingness to receive the COVID-19 vaccine.[27] Importantly, vaccine hesitancy may constitute a social identity for some,[28] which can strengthen opposition to vaccines.[29,30]

Alternatively, vaccine hesitancy may be grounded in community norms and practices. For example, many observant Muslims have foregone influenza vaccines that contained gelatin made from pork.[31] Similarly, distrust in vaccines and modern medicine can be influenced by concerns over historical abuses and atrocities committed by the medical establishment. For example, the Tuskegee experiments have translated into lowered trust in medicine and therefore decreased vaccine confidence among the African American community.[12,32] Ongoing racial discrimination within the health care system also contributes to decreased vaccine confidence among African Americans.[33] Similarly, political identity and rhetoric can also impact vaccine hesitancy and uptake. For example, in the United States and Italy, some political elites, such as populists and President Donald Trump, have advanced vaccine skepticism for political purposes even before the COVID-19 pandemic.[34] Politicization of vaccines can increase distrust toward medical experts among supporters of these politicians, though this politicization may also increase support for vaccines among those who oppose vaccine-skeptical politicians.[35] As a result, Republicans (vs Democrats) in the United States have been more likely to endorse vaccine misinformation and be vaccine hesitant.[36]

2.2 Influence of misinformation and conspiracy theories on vaccine hesitancy

Information, including misinformation and conspiracy theories, about vaccines plays a large role in shaping concerns related to vaccine hesitancy,[23,24,37,38] particularly during the COVID-19 pandemic.[39–45] For example, one content analysis of anti-vaccination websites found that conspiracy theories were present on every website analyzed, and that misinformation, such as the misrepresentation of vaccine studies, was extensive.[21]

Although misinformation and conspiracy theories about vaccines are not entirely new phenomena, the proliferation of the internet and social media has facilitated their spread and has entrenched belief.[46] Recent research suggests that access to the internet accounts for the greater levels of vaccine hesitancy in richer countries compared to poorer countries.[14] This is likely due to greater online access, which enables the spread of anti-vaccine misinformation.[21] Furthermore, the widespread availability of medical information provides the illusion that people who search and find information on vaccines via the internet are adequately and credibly informed, thus leading to higher confidence in their knowledge despite having wrong or incomplete information.[47]

Social media in particular allows users to share information broadly, at a fast pace, and with increased anonymity, features that allow users to quickly spread unverified information.[48] Indeed, evidence suggests that tweets that contain misinformation are retweeted faster and by more Twitter users on average than tweets that contain factual information.[49] This is also true of other social media platforms. A content analysis of YouTube videos about vaccination found that negative videos—including those that question vaccine science and those that promote conspiracy theories about collusion between vaccine supporters and pharmaceutical companies—were more likely to receive a rating, have higher ratings, and receive more views than positive videos (e.g., those describing the benefits and safety of vaccines).[50] An analysis of exposure to information about HPV vaccines on Twitter found that HPV vaccine coverage was lower in states where misinformation and conspiracies made up a higher proportion of exposures, indicating that vaccine misinformation and conspiracy theories on Twitter may reflect or influence vaccine acceptance.[45,51]

There is a well-established association between belief in conspiracy theories—specific to vaccines or otherwise—and anti-vaccine beliefs. One cross-sectional survey, conducted across 24 countries and Hong Kong, found that participants who believed more vaccine-related conspiracy theories were more likely to hold anti-vaccine attitudes. This association was significant across all locations, but was strongest in Western countries (US, Australia, UK, Canada, Germany).[23] Another study found that anti-vaccination beliefs are highly correlated with belief in the conspiracies that Obama is a Muslim and that the Bush administration knew about the 9/11 attacks ahead of time, indicating that anti-vaccination beliefs may be "best explained as an extension of a common psychological predisposition for conspiracy beliefs."[37] Beyond belief in specific conspiracy theories, respondents who report more conspiracist ideation—a general willingness to believe conspiracy theories—are more likely to report concerns about the MMR vaccine.[35] Additionally, recent research indicates that conspiratorial thinking—the predisposition to interpret events as the result of conspiracies—is an important predictor of belief in conspiracy theories and misinformation. Enders et al. found that the established association between social media use and belief in conspiracy theories is contingent on conspiratorial thinking, such that the association strengthens as conspiratorial thinking increases.[52]

Vaccine misinformation and conspiracy theories can directly impact vaccine uptake.[30,45] For example, belief in the conspiracy theory that the pharmaceutical industry has conspired to cover up evidence of the relationship between vaccines and autism can lead people to avoid vaccinating their children. Jolley and Douglas found that both endorsement of anti-vaccine conspiracy theories and exposure to material supporting anti-vaccine conspiracy theories is associated with lower intentions to vaccinate a fictitious child.[24]

2.3 Expected correlates of COVID-19 vaccine hesitancy

As our review above makes clear, the reasons for vaccine hesitancy can vary across different types of vaccines. The COVID-19 vaccine is no exception; several factors related to the development and dissemination of the COVID-19 vaccine have resulted in vaccine hesitancy unique to this vaccine.

For example, some are hesitant toward the COVID-19 vaccine because of its rapid development and dissemination; cross-nationally, many people believe that the COVID-19 vaccine was rushed and have been worried about the safety of a vaccine developed much quicker than usual.[53,54] Additionally, the COVID-19 vaccine has been overtly politicized. Even prior to the COVID-19 pandemic, Donald Trump became the first president to publicly share anti-vaccination attitudes.[35] In one study, exposure to President Trump's anti-vaccination tweets—often about the disproven link between the MMR vaccine and autism—increased concern about vaccines among Trump voters but not among liberal voters.[35] Once the COVID-19 pandemic ensued, the response was politicized during its early stages and before a vaccine was developed. For example, an analysis of tweets sent by members of Congress during the early months of the pandemic found that Democrats were more likely to tweet about COVID-19 earlier and more frequently, and their tweets were more likely to focus on public health and direct aid to workers, whereas Republicans' tweets focused more on the perceived role of the Chinese government.[55]

Several studies about indicate that willingness to receive a COVID-19 vaccine is lower among Republicans than among Democrats.[5,56] Additionally, a longitudinal study conducted during the first 6 months of the pandemic found that favorable attitudes toward vaccines and intention to be vaccinated for COVID-19 and the flu decreased over time, and that this decrease was driven by Republicans.[18] This political divide is reflected in vaccination rates as well. Democrats are more likely than Republicans to report having received the COVID-19 vaccine, and the share of the total population that is fully vaccinated has increased more rapidly in counties that voted for Biden than counties that voted for Trump.[57]

Belief in COVID-19 conspiracy theories, given the above discussion, is also likely to increase COVID-19 hesitancy.[45] Indeed, Allington et al. found that belief in COVID-19 conspiracy theories is predictive of COVID-19 vaccine hesitancy.[39] Another study also found that belief in conspiracy beliefs, related to COVID-19 specifically as well as vaccines more broadly, was positively associated with vaccine hesitancy.[42] Republicans are also more likely to believe COVID-19 conspiracy theories than Democrats,[58] which, as noted above, are associated with anti-vaccine beliefs. Relatedly, we expect conspiratorial thinking to also be associated with COVID-19 vaccine hesitancy. For instance, one study found that the gender gap in COVID-19 conspiracy belief was explained in part by differences in conspiratorial thinking.[59]

As noted above, vaccine hesitancy and vaccine misinformation endorsement tend to positively correspond with a distrust of experts and intellectuals,[27,60,61] which is often described as anti-intellectualism.[61] This concept is also deeply tied to more general populist attitudes. Stecuła and Pickup assess two of the core dimensions of populism (conflict between average citizens and elites, and distrust of experts and intellectuals), identifying that populist beliefs are correlated with conspiracy beliefs about COVID-19, above and beyond partisanship, particularly for those who consume conservative media.[85] Additionally in the context of COVID-19, work by Merkley and Loewen finds that higher respondent scores of anti-intellectualism negatively and significantly correspond with following scientifically-backed behavioral recommendations to prevent the spread of COVID-19.[62] This study also finds that anti-intellectualism significantly and positively predicts lower levels of COVID-19 information and news seeking behavior. Taken together, these studies suggest that anti-intellectualism should also be a significant predictor of COVID-19 vaccine hesitancy.

Given the above discussion, this study evaluates three explanations regarding vaccine hesitancy in the context of the COVID-19 pandemic: conspiracy theory beliefs (including both COVID-19 conspiracy theory beliefs and conspiratorial predispositions) political identity, and anti-intellectualism. Moreover, we explore whether any of these explanations is more robust than the others.

3. Hypotheses

Thus, we expect:

- Hypothesis 1a (H1a): Greater endorsement of COVID-19 conspiracy theories will covary with increased vaccine hesitancy and decreased vaccine support.
- Hypothesis 1b (H1b): Greater conspiratorial thinking will covary with increased vaccine hesitancy and decreased vaccine support.
- Hypothesis 2 (H2): That those on the political right will engage in greater vaccine hesitancy and lower reported vaccine support than those on the political left.
- Hypothesis 3 (H3): That those higher (vs lower) in anti-intellectualism will be more likely to be vaccine hesitant and less supportive of vaccines.

4. Data and methods

4.1 Data

We test these hypotheses and our research question using data from two panel studies, including one sampled to approximate national representativeness, recruited before and after the 2020 U.S. Presidential Election. The first sample was recruited from Amazon's Mechanical Turk platform (MTurk). MTurk samples are more diverse than student samples and more representative than typical Internet samples.[63,64] The second study was quota sampled by Forthright, an online research panel with national representative sampling capabilities, part of market research agency Bovitz, Inc. We use data from Wave 1 of both panel studies (Sample 1, October 23–30, 2020, $N = 1080$, 60.4% female and 39.5% male, 56.3% with at least a BA, 54.8% with family income greater than $50,000, mean age = 44.18, $SD = 14.66$; White = 83.98%; Sample 2, October 27–November 2, 2020, $N = 1140$, 52.80% female and 46.50% male, 32.6% with at least a BA, 50.7% with family income greater than $50,000, mean age = 46.34, $SD = 17.04$; White = 62.11%). Analyses are conducted using multivariate OLS regression through Stata 14.

4.2 Measures

All measures were rescaled to range from 0 to 1. Descriptive statistics for the two samples are provided within Table 1. Question wording and response scales are available in Appendix A. Replication data and syntax are available at:

4.3 Dependent variable

Vaccine support. Scholars have utilized multiple measures of vaccine hesitancy, ranging from a single item assessing vaccine intention[65,66] to more varied measures assessing multiple dimensions of vaccination.[67] Our dependent variable, based on Betsch et al.[67] is a measure of vaccine support and hesitancy. This measure is an index of five items scored on a one to seven agreement scale. This scale utilizes questions that encompass the dimensions of vaccine hesitancy described above, including confidence, complacency, and convenience. Respondents were asked items such as, "I am completely confident that vaccines are safe," and "When everyone else is vaccinated, I don't have to be vaccinated, too." The five items were averaged and recoded to range from 0 to 1 (Sample 1, mean = 0.71, std.dev = 0.19, alpha = 0.82; Sample 2, mean = 0.67, std.dev = 0.19, alpha = 0.79). Higher values correspond to greater vaccine support and less vaccine hesitancy; whereas lower values correspond to less vaccine support and more vaccine hesitancy.

4.4 Independent variables

Conspiracy theory belief. Our first explanatory variable is a COVID-19 CT index consisting of the average of responses to eight CT beliefs. The question wording for all measures are available in Appendix A. Prior work has established that COVID-19 conspiracy theories may be monological,[58] thus justifying the use of a composite index (Sample 1 alpha = 0.82, Sample 2 alpha = 0.80).

Conspiratorial thinking. Another dispositional factor that is positively related to CT beliefs (including endorsement of COVID-19 CTs)[58,68] is conspiratorial thinking—the tendency to view events as the product of a conspiracy.[69] However, as there may be an underlying predisposition toward conspiracy beliefs beyond endorsement of target-specific COVID-19 conspiracy theories which may also be subject to political motivated reasoning, we employ a more generalized measure of conspiracy theory belief that assesses an underlying predisposition for conspiracy theory belief. This was as measured via a four-item, five-point agree/disagree scale validated by

Table 1 Descriptive statistics.

		Sample 1		Sample 2	
		N	%	N	%
Age (in years)			Avg:		Avg:
	(Range S1 18–89; S2 18–79; S318–93)	1007	44.18	995	46.06
Sex					
	Female	650	60.24	515	52.18
	Male	429	39.76	472	47.82
Education					
	Up to and including high school diploma/GED	341	31.6	618	55.78
	Some post–high school, no bachelor's degree	519	48.2	381	34.38
	Bachelor's degree	82	7.6	90	8.12
	Graduate degree or post-bachelor's degree	402	29.2	19	1.71
Income					
	Under $19,999	118	10.9	431	38.41
	$20,000 to $29,999	108	10.0	208	18.54
	$30,000 to $49,999	259	21.7	304	28.88
	$50,000 to $99,999	398	36.9	159	14.17
	$100,000 and above	197	18.2	10	0.89
Race					
	Nonwhite	173	16.02	364	36.58
	White	907	83.98	631	63.42
Ideology					
	Extremely liberal	131	13.00	112	11.58
	Liberal	201	19.94	145	14.99
	Somewhat liberal	107	10.62	102	10.55
	Moderate	183	18.15	323	33.40
	Somewhat conservative	137	13.59	94	9.72
	Conservative	166	16.47	118	12.20

Continued

Table 1 Descriptive statistics.—cont'd

		Sample 1		Sample 2	
		N	%	N	%
	Extremely conservative	83	8.23	73	7.55
Party Identification					
	Republican (and leaners)	387	39.33	310	35.59
	Independents	80	8.13	113	12.97
	Democrats (and leaners)	517	52.54	448	51.44
COVID-19 CT Belief					
	T1 mean (std. dev)	1013	0.26 (0.20)	911	0.32 (0.21)
Conspiratorial Thinking					
	T1 mean (std. dev)	1007	0.48 (0.28)	909	0.52 (0.25)
Anti-intellectualism					
	T1 mean (std. dev)	542	0.28 (0.20)	487	0.31 (0.22)
Learned Helplessness					
	T1 mean (std. dev)	1010	0.33 (0.22)	917	0.36 (0.22)
External Efficacy					
	T1 mean (std. dev)	1009	0.36 (0.22)	910	0.40 (0.24)
Internal Efficacy					
	T1 mean (std. dev)	1009	0.35 (0.21)	910	0.39 (0.22)
Political Interest					
	T1 mean (std. dev)	1010	0.69 (0.22)	908	0.63 (0.24)
Democratic Satisfaction					
	T1 mean (std. dev)	1004	0.52 (0.20)	915	0.52 (0.21)
Trust					
	T1 mean (std. dev)	1009	0.43 (0.17)	909	0.42 (0.18)
Political Knowledge					
	T1 mean (std. dev)	1005	0.81 (0.23)	916	0.70 (0.27)

Note: Sample 1: MTurk; Sample 2: Forthright.

Uscinski et al.[70] The measures are internally consistent across samples (Sample 1 alpha = 0.88; Sample 2 alpha = 0.84).

Political identity. As political predispositions have been widely associated with beliefs in COVID-19 related conspiracy theories[58] as well as likelihood to vaccinate,[5,18,36,56] we controlled for an *ideological self-placement* and *partisanship*. Ideology was measured using a seven-point measure ranging from extremely liberal = 1 to extremely conservative = 7. Partisanship was assessed using two dummy variables for Republican and Independent, with Democratic identity serving as the reference category.

Anti-intellectualism. In recent years, scholars have expanded their investigation of anti-intellectualism, linking it not only to political outcomes and support for politicians and political movements that are skeptical of experts,[61] split by partisan identities,[71] but also to the COVID-19 pandemic.[62] In order to assess anti-intellectualism, we utilized the three-items specifically aimed at measuring anti-intellectualism borrowed from a larger, 20-item measure developed by Oliver and Rahn, also utilized by Stecuła and Pickup (Sample 1 alpha = 0.69; Sample 2 alpha = 0.68).[72,85]

Control variables. Prior work has demonstrated that those who are politically disaffected are distrusting of the political system, less likely to engage in politics, less satisfied with political systems, and tend to engage in greater conspiratorial thinking.[73–75] Given that the latest vaccines have been supported by government institutions, and because many communities have been marginalized and experienced abusive relationships with mandatory vaccines we control for five measures of political disaffection.[76] *Trust* was measured using a generalized index of trust in federal and local government, media, people in general, and law enforcement (Sample 1 alpha = 0.68, Sample 2 alpha = 0.72). *Internal and External Efficacy* were each measured using short two-item batteries,[77] both presenting moderate consistent internal consistency across samples (IE: Sample 1 alpha = 0.54, Sample 2 alpha = 0.44; EE: Sample 1 alpha = 0.70, Sample 2 alpha = 0.67). While similar to some measures of efficacy, we also assessed *Learned Helplessness,* which provides a unique approach to assess perceptions of loss of control due to one's repeated attempts and failures, inside and outside of the political domain.[78] People who are higher in learned helplessness (LH) are more likely to engage in conspiratorial thinking[74] and more likely to engage in COVID-19 related conspiracy theories.[59] It was measured here using a short six-item battery validated by Farhart,[73] which was taken from the original 20-items.[79] This index also presented substantial internal consistency across the two samples (Sample 1 alpha = 0.90, Sample 2 alpha = 0.87). *Satisfaction* was assessed using a two-item index averaging satisfaction with democracy and life in general (Sample 1 alpha = 0.65, Sample 2 alpha = 0.65).[80] All measures were recoded to scale from 0 to 1, with higher values correspond to higher trust, efficacy, helplessness, and satisfaction.

We also include other demographic and attitudinal control variables as covariates. First, we included the following demographics: *age* (rescaled 0–1), *income* (ordered categories rescaled from 0 to 1), *gender* (0 = male, 1 = female), *education* (ordered categories rescaled from 0 to 1), and *race* (0 = nonwhite, 1 = white). Second, we controlled for knowledge about and interest in politics using two variables (1) *Political knowledge* was assessed using eight factual political knowledge items (Sample 1 alpha = 0.76, Sample 2 alpha = 0.76).[81] Items were scored 0 (incorrect) or 1 (correct), summed, and averaged; (2) P*olitical interest* was assessed using three items that evaluated interest in and importance of politics (Sample 1 alpha = 0.75, Sample 2 alpha = 0.75).[82]

5. Results

5.1 Effect of conspiracy belief on vaccine support and hesitancy

Given extant literature, we expect that greater conspiracy belief would lead to greater vaccine hesitancy and reduced vaccine likelihood. To test this in a more robust way, we examine our first two hypotheses with two separate measures of conspiracy belief: an index of COVID-19 related conspiracy theory beliefs (H1a) and a predisposition toward conspiratorial ideation, conspiratorial thinking (H1b). We find that across both samples and for both measures of conspiracy belief reported in Table 2 that indeed greater conspiracy belief leads to greater vaccine hesitancy demonstrated by reduced vaccine support (COVID-19 CT index: Model 1 $b = -0.56$, $P < 0.05$; Model 2 $b = -0.45$, $P < 0.05$; conspiratorial thinking: Model 3 $b = -0.27$, $P < 0.05$; Model 4 $b = -0.25$, $P < 0.05$). Through four separate tests of the effect of conspiracy theory belief on vaccine hesitancy, we find strong support for both H1a and H1b.

Fig. 1 demonstrates the relative effect of each covariate on vaccine hesitancy illustrated by vaccine support through the display of coefficients across models by sample.

Table 2 Effects of conspiracy theory belief, political identity, and anti-intellectualism on vaccine support.

	Sample 1	Sample 2	Sample 1	Sample 2	Sample 1	Sample 2	Sample 1	Sample 2	Sample 1	Sample 2
	(1)	**(2)**	**(3)**	**(4)**	**(5)**	**(6)**	**(7)**	**(8)**	**(9)**	**(10)**
Age	−0.05	−0.06	−0.03	−0.07 +	0.01	−0.05	−0.00	−0.06	−0.04	−0.06
	(0.03)	(0.03)	(0.03)	(0.04)	(0.04)	(0.04)	(0.05)	(0.05)	(0.05)	(0.05)
Female	−0.00	0.01	−0.00	0.00	−0.00	−0.01	0.00	−0.03 +	−0.00	−0.01
Dummy	(0.01)	(0.01)	(0.01)	(0.02)	(0.02)	(0.02)	(0.02)	(0.02)	(0.02)	(0.02)
White	0.01	0.02	0.03	0.03	0.06*	0.07*	0.02	0.02	0.00	0.03
Dummy	(0.02)	(0.02)	(0.02)	(0.02)	(0.03)	(0.02)	(0.03)	(0.02)	(0.04)	(0.03)
Income	0.01	0.01*	0.05*	0.01	0.04	0.01 +	0.03	0.01	−0.03	0.01*
	(0.02)	(0.00)	(0.03)	(0.00)	(0.03)	(0.00)	(0.03)	(0.00)	(0.04)	(0.00)
Education	−0.02	0.05	−0.02	0.08*	0.00	0.07 +	−0.07	−0.00	−0.06	−0.06
	(0.03)	(0.03)	(0.04)	(0.03)	(0.04)	(0.04)	(0.05)	(0.04)	(0.06)	(0.05)
Political	0.00	0.05	0.11*	0.15*	0.12*	0.17*	0.12*	0.15*	0.00	0.07
Knowledge	(0.03)	(0.03)	(0.04)	(0.03)	(0.04)	(0.04)	(0.05)	(0.05)	(0.05)	(0.05)
Political	0.06*	0.00	0.04	0.01	−0.03	0.00	0.03	−0.03	0.02	−0.02
Interest	(0.03)	(0.04)	(0.04)	(0.04)	(0.04)	(0.04)	(0.05)	(0.05)	(0.05)	(0.05)
Democratic	−0.07 +	−0.05	−0.13*	−0.11*	−0.07	−0.07	0.06	−0.11 +	0.04	−0.05

Continued

Table 2 Effects of conspiracy theory belief, political identity, and anti-intellectualism on vaccine support.—cont'd

	Sample 1	Sample 2	Sample 1	Sample 2	Sample 1	Sample 2	Sample 1	Sample 2	Sample 1	Sample 2
	(1)	(2)	(3)	(4)	(5)	(6)	(7)	(8)	(9)	(10)
Satisfaction	(0.04)	(0.05)	(0.04)	(0.04)	(0.05)	(0.05)	(0.06)	(0.06)	(0.06)	(0.07)
Internal	−0.02	−0.02	−0.05	0.03	−0.11*	0.05	−0.09	0.04	−0.04	0.05
Efficacy	(0.04)	(0.04)	(0.05)	(0.04)	(0.05)	(0.05)	(0.06)	(0.06)	(0.06)	(0.06)
External	−0.03	−0.04	−0.06	−0.07 +	−0.04	−0.09*	−0.06	−0.01	−0.05	−0.16*
Efficacy	(0.03)	(0.04)	(0.04)	(0.04)	(0.04)	(0.04)	(0.06)	(0.05)	(0.06)	(0.07)
Trust	−0.05	0.02	−0.07	0.01	0.01	0.08	0.01	0.10	−0.02	0.05
	(0.05)	(0.05)	(0.05)	(0.05)	(0.05)	(0.06)	(0.07)	(0.07)	(0.07)	(0.08)
Learned	−0.13*	−0.04	−0.14*	−0.10*	−0.15*	−0.18*	−0.03	−0.10 +	−0.04	−0.08
Helplessness	(0.03)	(0.05)	(0.04)	(0.05)	(0.04)	(0.05)	(0.05)	(0.06)	(0.06)	(0.06)
COVID CT	−0.56*	−0.45*							−0.39*	−0.38*
Index	(0.04)	(0.05)							(0.10)	(0.08)
Conspiratorial			−0.27*	−0.25*					−0.13*	−0.09
Thinking			(0.02)	(0.04)					(0.05)	(0.06)
Ideology					−0.21*	−0.11*			−0.06	0.00
					(0.04)	(0.04)			(0.06)	(0.06)

Republican					0.01	−0.09*			0.02	−0.08*
Dummy					(0.02)	(0.03)			(0.04)	(0.04)
Independent					0.02	−0.07*			−0.03	−0.15*
Dummy					(0.03)	(0.03)			(0.04)	(0.05)
Anti-							−0.44*	−0.37*	−0.19*	−0.18*
Intellectualism							(0.05)	(0.05)	(0.08)	(0.07)
Constant	0.94*	0.82*	0.88*	0.77*	0.78*	0.68*	0.76*	0.77*	1.02*	1.01*
	(0.05)	(0.06)	(0.06)	(0.06)	(0.06)	(0.06)	(0.08)	(0.07)	(0.09)	(0.08)
N	570	477	567	477	503	420	312	255	207	170
adj. R2	0.434	0.320	0.312	0.224	0.295	0.233	0.313	0.290	0.524	0.493

Note: Sample 1: MTurk; Sample 2: Forthright. Standard errors in parentheses + $P < 0.10$, * $P < 0.05$.

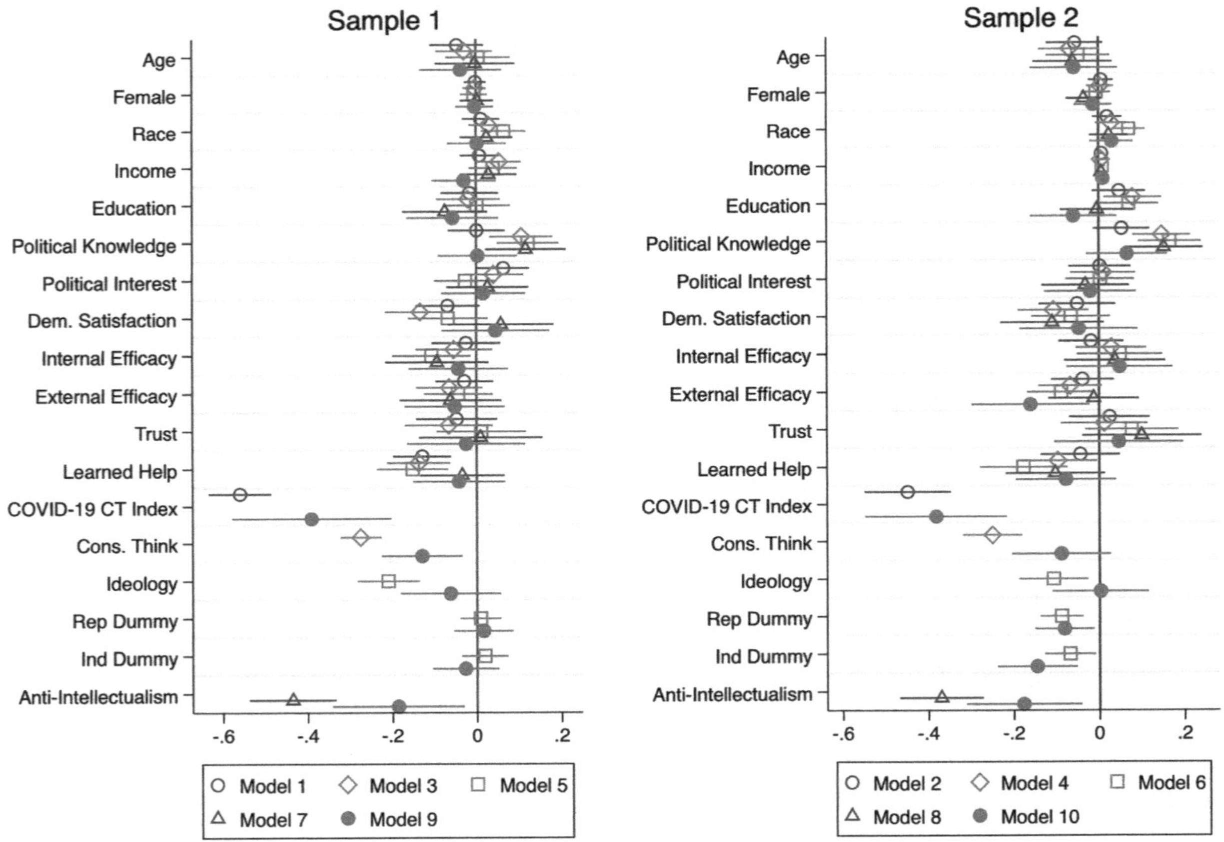

Fig. 1 Coefficient plot illustrating main effects on vaccine support.

5.2 Effect of political identity on vaccine support and hesitancy

Next, we assess the effect of political identity on vaccine hesitancy. As a result, we expect that those on the political right will exhibit greater vaccine hesitancy. We test this hypothesis (H2) by examining the effect of political ideology and partisanship on vaccine hesitancy. Across both samples reported in Table 2, we find robust evidence in support for political ideology such that as conservativism increases, vaccine hesitancy increases and support significantly decreases (Model 5 $b = -0.21$, $P < 0.05$; Model 6 $b = -0.11$, $P < 0.05$). However, the results for partisanship are more mixed across samples. In Sample 1, we do not see a significant effect of partisanship, controlling for political ideology, on vaccine support. In contrast, Sample 2 shows that both Republicans and Independents, relative to Democrats (and while controlling for political ideology), significantly report reduced vaccine support (Model 6 Republicans $b = -0.09$, $P < 0.05$; Independents $b = -0.07$, $P < 0.05$). Thus, in three out of four tests of the effect of political identity on vaccine support and hesitancy, we find evidence in support of H2.

5.3 Effect of anti-intellectualism on vaccine support and hesitancy

Our third hypothesis engages the question of whether anti-intellectualism is robustly related to vaccine hesitancy. We expect that those who report higher levels of anti-intellectualism will be more vaccine hesitant (H3). Across both samples, we find strong effects of anti-intellectualism on increased vaccine hesitancy illustrated through significant decreases in vaccine support (Model 7 $b = -0.44$, $P = 0.05$; Model 8 $b = -0.37$, $P < 0.05$; Table 2). This illustrates that there are additional concerns for the way in which decreased trust in experts and science can contribute to potentially dire public health consequences such as increased vaccine hesitancy and potentially reduced uptake, particularly as the COVID-19 vaccine has been identified as one of the most robust mechanisms for curbing the pandemic and slowing the spread of COVID-19.

5.4 Comparative effects

While testing each of the prior hypotheses independently provides support for our expectations, we are unable to assess them against one another. Thus, we ask a more general research question as to whether one explanation—conspiracy belief, political identity, or anti-intellectualism—is more robust in explaining vaccine hesitancy during the COVID-19 pandemic. To evaluate this research question, we included all of our key measures in the same model to assess covariation and multicollinearity. For Samples 1 and 2 reported in Models 9 and 10, we find that the COVID-19 CT index is the most robust predictor, such that as COVID-19 conspiracy belief

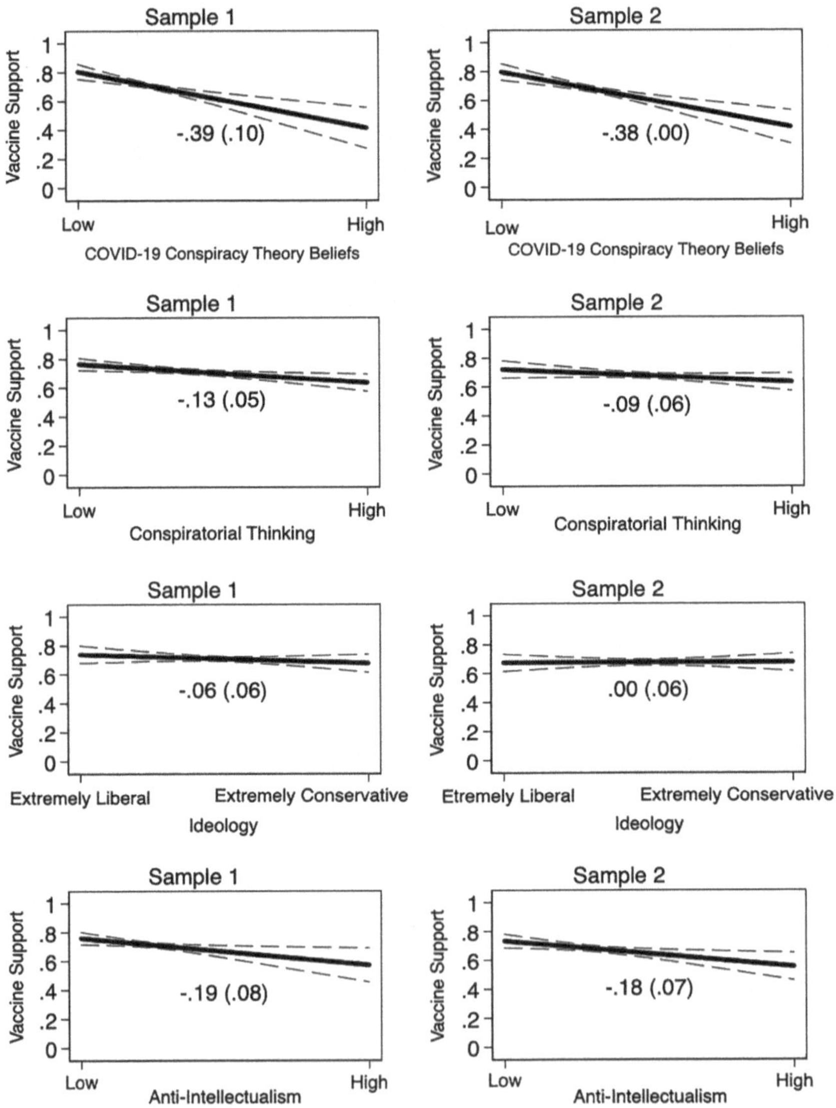

Fig. 2 Effects of conspiracy belief, political identity, and anti-intellectualism on vaccine Support.

increases, vaccine hesitancy clearly increases illustrated by significant decreases in vaccine support (Model 8 $b = -0.39$, $P < 0.05$; Model 9 $b = -0.38$, $P < 0.05$). Moving from low conspiracy belief to high conspiracy belief, we see a 38% and 39% increase in vaccine hesitancy. Conspiratorial thinking is still statistically significant in Sample 1 (Model 8 $b = -0.13$, $P < 0.05$), but drops out of significance in Sample 2. This may suggest that specific COVID-19 conspiracy beliefs, rather than general conspiratorial orientations, are more consequential for vaccine hesitancy. Again, in these comparative models, we find inconsistent effects of political identity. Political ideology drops out of significance in both samples, and partisanship is only significant in Sample 2 (Model 10 Republicans $b = -0.08$, $P < 0.05$; Independents $b = -0.15$, $P < 0.05$). Lastly, while not quite as large but still just as robust, anti-intellectualism remains a strong predictor for increased vaccine hesitancy across both samples (Model 9 $b = -0.19$, $P < 0.05$; Model 10 $b = -0.18$, $P < 0.05$). This carries substantive effects such that moving from low anti-intellectualism to high anti-intellectualism translates to an 18% and 19% increase in vaccine hesitancy. Fig. 2 also demonstrates the independent effects for each key measure for vaccine hesitancy for Models 9 and 10 across the samples.

6. Discussion and conclusion

In this study, we sought to assess whether three prominent explanations for vaccine hesitancy are also affecting vaccine uptake in the context of the COVID-19 pandemic. We find that COVID-19 related conspiracy belief and anti-intellectualism have the most consistent effects on vaccine hesitancy in the COVID-19 context, in line with expectations from other research.[39,45,62] Additionally, political identity—assessed through both political ideology and partisanship—was also consequential, consistent with prior research.[5,56] Specifically, self-identified conservatism and identification with the Republican (vs Democratic) party covaried with reduced vaccine uptake, although these effects were no longer significant in three of four tests when accounting for conspiratorial thinking, conspiracy theories, and anti-intellectualism in the same models. This stands in contrast to some previous work suggesting that vaccine hesitancy, including COVID-19 vaccine hesitancy, corresponds strongly with partisan identities.[5,18,36,56,58] This finding is likely due to other significant factors, such as conspiracy theory belief[83] or epistemic hubris and anti-intellectualism,[71] being closely related to partisanship and political ideology. Together, these results demonstrate that attitudes toward expertise, belief in COVID-19 conspiracies theories, and (to a much lesser extent) identification with conservative partisan and ideological groups undermine support for vaccination, independent of a host of control variables known to predict attitudes toward science, perceptions of the pandemic, and willingness to engage in COVID-19 behavioral mitigation.

One key limitation of this study is that both samples were conducted in the US context, where polarization on policy responses and rampant spread of misinformation and conspiracy theories about the virus have taken place. Previous research has found that base levels of general vaccine hesitancy vary widely,[14] as well as levels of COVID-19 vaccine hesitancy and uptake.[6,27] Taken together, non-US contexts may yield findings different from those in the present study. Additionally, the data was collected before any vaccines were granted emergency use authorization through the FDA, and attitudes have continued to evolve following the approval of multiple vaccines and large campaigns to improve vaccine uptake across the United States and worldwide. However, vaccine hesitancy remains an obstacle for mass vaccination efforts.

The high degree of politicization and polarization about the virus, particularly at the beginning of the COVID-19 pandemic, added to the influence of right-learning media regularly discussing CTs and misinformation, led audiences to see COVID-19 as less dangerous and negatively impacted engagement with protective behaviors and vaccine support (e.g., Motta, Stecuła, Farhart 2020; Romer and Hall Jamieson 2020; Romer and Hall Jamieson 2021),[38,86,87] and even more significantly in the United States than the United Kingdom (Pennycook et al. 2020a).[88] The pandemic has activated a perfect storm of psychological, political, and situational factors at the root of conspiracy theory beliefs, particularly those emerging regarding COVID-19.[58] These specific conspiracy theory beliefs have challenged governments, and health and medical experts in their journey to control the spread of the virus and tackle resistance to preventative and containment-related behaviors,[89,90] as well as future vaccine likelihood.[86]

Combatting science skepticism, medical mistrust, anti-intellectualism, and conspiracy beliefs, while working toward increasing vaccine uptake is exceptionally challenging. Health officials could more frequently discuss with the public the risks of the virus, mechanisms of transmission, and confront science skepticism to reinforce the integrity of the scientists working to gather data and combat infectious diseases, possibly through fact checks.[91] Over the course of the COVID-19 pandemic, scholars have sought solutions and suggestions for policymakers and public health officials. One solution may lie with developing work on conspiracy theory inoculation.[93,94,99,100] Specifically in the context of COVID-19, tracking COVID-19 conspiracy theories and vaccine misinformation in real-time and engaging with social media to disseminate correct information may inoculate and help safeguard the public against misinformation and misinformed behaviors.[95] Framing of messaging aimed at targeting vaccine hesitancy is essential—individuals who receive information about the safety and efficacy of the COVID-19 vaccine may be more open to receiving the vaccine, whereas those who receive information that others are vaccine reluctant are going to report greater vaccine hesitancy themselves.[44] Moreover, urging social media users to consider the accuracy of the information they share has the potential to reduce the spread of COVID-19 conspiracy theories and misinformation.[96] Additionally, encouraging collectivism in messaging could increase engagement with protective behaviors.[97,98] However, given prior experience with Zika and yellow fever, corrective information may not entirely reduce misperceptions or improve support for control policies and intentions to engage in preventive behaviors.[92]

The results of our study raise important questions about the spread of conspiracy theories and misinformation about COVID-19, the growth of anti-intellectualism, and the consequences for public health, particularly in regard to attacks against the vaccine.[84] As national and international health agencies and governments continue to struggle to respond to the global spread of COVID-19, they are faced with the spread of misinformation and conspiracy theories about the virus. Lower-quality information environments, along with growing anti-intellectualism and entrenched political polarization, undermines public debate and understanding surrounding protective behaviors and policies in the context of the COVID-19 pandemic. In particular, our results demonstrate that on-going mass vaccination efforts may need to navigate the landmines created by anti-intellectual attitudes, vaccine misinformation, and COVID-19 conspiracy theories more acutely. This is no small matter. It is a critical priority for research going forward to inform the way in which pandemic responses should be formulated and communicated, vaccination efforts should be facilitated and sustained, and how we can curb the continued spread of COVID-19 and the pernicious consequences of misinformation worldwide.

Appendix A. Question Wording

(Response options in italics; questions are identical across surveys unless otherwise noted)

A.1 Vaccine hesitancy

I am completely confident that vaccines are safe.

Vaccination is unnecessary because vaccine-preventable diseases are not common anymore.

Everyday stress prevents me from being vaccinated.

Vaccination is unnecessary because vaccine-preventable diseases are not common anymore.

Everyday stress prevents me from being vaccinated.

When I think about being vaccinated, I weigh its benefits and risks to make the best decision possible.

When everyone else is vaccinated, I dont have to be vaccinated, too.

Response options: *Strongly disagree, Disagree, Somewhat disagree, Neither agree nor disagree, Somewhat agree, Agree, Strongly agree.*

A.2 COVID-19 conspiracy theory belief

Some people believe that the coronavirus (COVID-19) is a bioweapon engineered by the Chinese government to wage war on America and Western countries. Others do not believe this. What do you think? The coronavirus (COVID-19) is…

Definitely a bioweapon engineered by the Chinese government to wage war.

Probably a bioweapon engineered by the Chinese government to wage war.

Probably not a bioweapon engineered by the Chinese government to wage war.

Definitely not a bioweapon engineered by the Chinese government to wage war.

Some people believe that the coronavirus (COVID-19) panic has been an orchestrated effort by powerful people to close down businesses and destroy capitalism. Others do not believe this. What do you think? The coronavirus (COVID-19) panic…

Definitely has been orchestrated by powerful people to close down businesses and destroy capitalism.

Probably has been orchestrated by powerful people to close down businesses and destroy capitalism.

Probably has not been orchestrated by powerful people to close down businesses and destroy capitalism.

Definitely has not been orchestrated by powerful people to close down businesses and destroy capitalism.

Some people believe that the coronavirus (COVID-19) was originally engineered by the U.S. military. Others do not believe this. What do you think? The coronavirus (COVID-19) was…

Definitely engineered by the U.S. military.

Probably engineered by the U.S. military.

Probably not engineered by the U.S. military.

Definitely not engineered by the U.S. military.

Some people believe that the infection rate from coronavirus (COVID-19) is different than is reported, so as to cover up how many people have been infected. Others do not believe this. What do you think? The coronavirus (COVID-19) infection rate is…

Definitely different than reported to cover up the number of infections.

Probably different than reported to cover up the number of infections.

Probably not different than reported to cover up the number of infections.

Definitely not different than reported to cover up the number of infections.

Some people believe that former Microsoft CEO Bill Gates is creating a tracking device to be injected with the coronavirus vaccine. Others do not believe this. What do you think? Former Microsoft CEO Bill Gates…

Definitely is creating a tracking device to be injected with the coronavirus vaccine.

Probably is creating a tracking device to be injected with the coronavirus vaccine.

Probably is not creating a tracking device to be injected with the coronavirus vaccine.

Definitely is not creating a tracking device to be injected with the coronavirus vaccine.

Some people believe that 5G technology is causing the coronavirus to spread faster. Others do not believe this. What do you think? 5G technology …

Definitely is causing the coronavirus to spread faster.

Probably is causing the coronavirus to spread faster.

Probably is not causing the coronavirus to spread faster.

Definitely is not causing the coronavirus to spread faster.

Some people believe that Donald Trump is lying about getting COVID-19 in order to improve his reelection prospects. Others do not believe this. What do you think? Donald Trump…

Definitely is lying about getting COVID-19 in order to improve his reelection prospects.

Probably is lying about getting COVID-19 in order to improve his reelection prospects.

Probably is not lying about getting COVID-19 in order to improve his reelection prospects.

Definitely is not lying about getting COVID-19 in order to improve his reelection prospects.

Some people believe that Donald Trumps political opponents intentionally infected him with COVID-19 in order to undermine his reelection prospects. Others do not believe this. What do you think? Donald Trumps opponents…

Definitely infected him intentionally with COVID-19 in order to undermine his reelection prospects.

Probably infected him intentionally with COVID-19 in order to undermine his reelection prospects.

Probably did not infected him intentionally with COVID-19 in order to undermine his reelection prospects.

Definitely did not infected him intentionally with COVID-19 in order to undermine his reelection prospects.

A.3 Conspiracy predisposition/conspiratorial thinking

Much of our lives are being controlled by plots hatched in secret places.

Even though we live in a democracy, a few people will always run things anyway.

The people who really run the country are not known to the voters.

Big events like wars, economic recessions, and the outcomes of elections are controlled by small groups of people who are working in secret against the rest of us.

Response options: *Strongly disagree, Somewhat disagree, Neither agree nor disagree, Somewhat agree, Strongly agree.*

A.4 Partisanship

Generally speaking, do you usually think of yourself as a Democrat, a Republican, an Independent, or what? *Democrat, Republican, Independent, Other Party, please specify.*

[Branched if Democrat] Would you call yourself a strong Democrat or a not very strong Democrat? *Strong Democrat, Not very strong Democrat.*

[Branched if Republican] Would you call yourself a strong Republican or a not very strong Republican? *Strong Republican, Not very strong Republican*

[Branched if Independent or Other Party] Do you think of yourself as closer to the Democratic Party or the Republican Party? *Closer to the Democratic Party, Closer to the Republican Party, Closer to Neither Party*

A.5 Political ideology

We hear a lot of talk these days about liberals and conservatives. Here is a seven-point scale on which the political views that people might hold are arranged from extremely liberal to extremely conservative. Where would you place yourself on this scale? *Very Liberal, Somewhat Liberal, Middle of the Road, Somewhat Conservative, Conservative, Very Conservative.*

A.6 Anti-intellectualism

Id rather put my trust in the wisdom of ordinary people than the opinions of experts and intellectuals.

When it comes to really important questions, scientific facts dont help.

Ordinary people can really use the help of experts to understand complicated things like science and health.

Response Options: *Strongly agree, Agree, Somewhat agree, Neither agree nor disagree, Somewhat disagree, Disagree, Strongly disagree.*

A.7 Trust

How much of the time do you think you can trust the following to do what is right:

The federal government in Washington, D.C.

Local government.

Law enforcement.

Media.

People in general.

Response Options: *Almost always, Most of the time, Some of the time, Almost never.*

A.8 Internal efficacy

How often do politics and government seem so complicated that you cant really understand whats going on?

Always, Most of the time, About half the time, Some of the time, Never.

How well do you understand the important political issues facing our country?

Extremely well, Very well, Moderately well, Slightly Well, Not well at all.

A.9 External efficacy

How much do public officials care about what people like you think?

A great deal, A lot, A moderate amount, A little, Not at all.

How much can people like you affect what the government does?

A great deal, A lot, A moderate amount, A little, Not at all.

A.10 Learned helplessness

No matter how much energy I put into a task, I feel I have no control over the outcome.

Other people have more control over their success and/or failure than I do.

I feel that I have little control over the outcomes of my work.

I feel that anyone else could be better than me in most tasks.

No matter how hard I try, things never seem to work out the way I want them to.

When I do not succeed at a task, I do not attempt any similar tasks because I feel that I will fail them also.

Response Options: *Strongly agree, Agree, Disagree, Strongly disagree.*

A.11 Satisfaction

On the whole, how satisfied are you with the way democracy works in the United States?

All things considered, how satisfied are you with your life as a whole these days?

Response Options: *Very satisfied, Fairly satisfied, Not very satisfied, Not at all satisfied.*

A.12 Political knowledge

We would like to ask you a few questions about public figures and the political system in general. Please respond to each of the following questions as thoroughly as possible. Please do NOT look up answers to these questions online. It is very important that you provide your honest answer.

What job or political office does John Roberts currently hold? *Secretary of Defense, Attorney General, Senate Majority Leader, Secretary of the Interior, Justice of the Supreme Court.*

What job or political office does Mike Pence currently hold?

Attorney General, Vice President, Secretary of State, Speaker of the House, Governor of New Hampshire.

What job or political office does Boris Johnson currently hold? *Speaker of the United Nations General Assembly, Prime Minister of the United Kingdom, Minister of Australia, U.S. envoy to the United Nations, Head of the European Commission.*

What job or political office does Nancy Pelosi currently hold?

Attorney General, Vice President, Secretary of State, Speaker of the House, House Chief of Staff.

Which political party currently has the most members in the Senate in Washington?

Democrats, Republicans, Both parties have the same number of members.

How long is the term of office for a U.S. senator?

2 years, 4 years, 5 years, 6 years, 8 years.

Which political party currently has the most members in the House of Representatives in Washington? *Democrats, Republicans, Both parties have the same number of members.*

Whose responsibility is it to nominate judges to the Federal Courts — the President, the Congress, or the Supreme Court? *The President, Congress, The Supreme Court.*

A.13 Political interest

Some people seem to follow whats going on in government and public affairs most of the time, whether theres an election going on or not. Others arent that interested. Would you say you follow whats going on in government and public affairs most of the time, some of the time, only now and then, or hardly at all?

Most of the time, Some of the time, Only now and then, Hardly at all.

My political attitudes and beliefs are an important reflection of who I am.

Strongly disagree, Disagree, Somewhat disagree, Neither agree nor disagree, Somewhat agree,

Agree, Strongly agree.

In general, my political attitudes and beliefs are an important part of my self-image.

Strongly disagree, Disagree, Somewhat disagree, Neither agree nor disagree, Somewhat agree,

Agree, Strongly agree.

A.14 Age

What age did you turn on your last birthday?

A.15 Race

Please indicate your race/ethnicity. [Select all that apply].

Latino/Hispanic, Black/African American, Asian/Asian American, White/ Caucasian, Native American, Other Self-identify[].

A.16 Gender

What is your gender? *Male, Female, Self-identify[].*

A.17 Education

What is your highest level of education?

elementary school, junior high, some high school, high school graduate, some college, associates degree, bachelors degree, some graduate school, masters degree, MBA, JD, MD, PhD, other advanced degree.

A.18 Income

What is your total family (including parent income if dependent on parents) income?

Less than $10,000, $10,000–$19,999, $20,000–$29,999, $30,000–$39,999, $40,000–$49,999, $50,000–$59,999, $60,000–$69,999, $70,000–$79,999, $80,000–$89,999, $90,000–$99,999, $100,000 or greater.

References

1. Schwartz JL. New media, old messages: Themes in the history of vaccine hesitancy and refusal. *AMA J Ethics*. 2012;14(1):50–55.
2. Joslyn MR, Sylvester SM. The determinants and consequences of accurate beliefs about childhood vaccinations. *Am Politics Res*. 2017;47(3):628–649. https://doi.org/10.1177/1532673X17745342.
3. Stecula D, Kuru O, Albarracin D, Jamieson KH. Policy views and negative beliefs about vaccines in the United States, 2019. *Am J Public Health*. 2020;110(10):1561–1563. https://doi.org/10.2105/AJPH.2020.305828.
4. World Health Organization. *Ten threats to global health in 2019*; 2019. Accessed September 15, 2021 https://www.who.int/emergencies/ten-threats-to-global-health-in2019.
5. Callaghan T, Moghtaderi A, Lueck JA, et al. Correlates and disparities of intention to vaccinate against COVID-19. *Soc Sci Med*. 2021;113638.
6. Sallam M. COVID-19 vaccine hesitancy worldwide: A concise systematic review of vaccine acceptance rates. *Vaccine*. 2021;9(2):160. https://doi.org/10.3390/vaccines9020160.
7. Mandavilli A. *The Delta variant is the symptom of a bigger threat: Vaccine refusal. The New York Times*; 2021. https://www.nytimes.com/2021/07/25/health/coronavirus-vaccine-refusal.html. Accessed 15 September 2021.
8. MacDonald NE. Vaccine hesitancy: Definition, scope and determinants. *Vaccine*. 2015;33(34):4161–4614. https://doi.org/10.1016/j.vaccine.2015.04.036.
9. Salmon DA, Dudley MZ, Glanz JM, Omer SB. Vaccine hesitancy: Causes, consequences, and a call to action. *Vaccine*. 2015;33(4):D66–D71.
10. Troiano G, Nardi A. Vaccine hesitancy in the era of COVID-19. *Public Health*. 2021;194:245–251. https://doi.org/10.1016/j.puhe.2021.02.025.
11. Murthy BP, Sterrett N, Weller D, et al. Disparities in COVID-19 vaccination coverage between urban and rural counties—United States, December 14, 2020–April 10, 2021. *Morb Mortal Wkly Rep*. 2021;70(20):759.

12. Quinn S, Jamison A, Musa D, Hilyard K, Freimuth V. Exploring the continuum of vaccine hesitancy between African American and white adults: Results of a qualitative study. *PLoS Currents*. 2016;8.
13. Gallup. *Wellcome Global Monitor Report: How Does the World Feel about Science and Health?* 2019 [Accessed September 15, 2021. https://wellcome.ac.uk/sites/default/files/wellcome-global-monitor-2018.pdf].
14. Lunz Trujillo K, Motta M. How internet access drives global vaccine skepticism. *Int J Public Opin Res*. 2021. https://doi.org/10.1093/ijpor/edab012.
15. Bauch CT, Earn DJ. Vaccination and the theory of games. *Proc Natl Acad Sci U S A*. 2004;101(36):13391–13394. https://doi.org/10.1073/pnas.0403823101.
16. Schwarzer R, Fuchs R. Self-efficacy and health behaviours. In: Conner M, ed. *Predicting health behaviour: Research and practice with social cognition models*. Open University Press; 1996.
17. Zitek EM, Schlund RJ. Psychological entitlement predicts noncompliance with the health guidelines of the COVID-19 pandemic. *Pers Individ Differ*. 2021;2021(171): 110491.
18. Fridman A, Gershon R, Gneezy A. COVID-19 and vaccine hesitancy: A longitudinal study. *PLoS One*. 2021;16(4): e0250123.
19. Dubé E, Vivion M, MacDonald NE. Vaccine hesitancy, vaccine refusal and the anti-vaccine movement: Influence, impact and implications. *Expert Rev Vaccines*. 2015;14(1):99–117.
20. Reno C, Maietti E, Fantini MP, et al. Enhancing COVID-19 vaccines acceptance: Results from a survey on vaccine hesitancy in northern Italy. *Vaccine*. 2021;9(4):378.
21. Kata A. A postmodern Pandora's box: Anti-vaccination misinformation on the internet. *Vaccine*. 2010;28(7):1709–1716. https://doi.org/10.1016/j.vaccine.2009.12.022.
22. Callaghan T, Motta M, Sylvester S, Trujillo KL, Blackburn CC. Parent psychology and the decision to delay childhood vaccination. *Soc Sci Med*. 2019;238, 112407.
23. Hornsey MJ, Harris EA, Fielding KS. The psychological roots of anti-vaccination attitudes: A 24-nation investigation. *Health Psychol*. 2018;37(4):307. https://doi.org/10.1037/hea0000586.
24. Jolley D, Douglas KM. The effects of anti-vaccine conspiracy theories on vaccination intentions. *PLoS One*. 2014;9(2): e89177. https://doi.org/10.1371/journal.pone.0089177.
25. Freeman D, Lambe S, Yu L, et al. Injection fears and COVID-19 vaccine hesitancy. *Psychol Med*. 2021;1–11. https://doi.org/10.1017/S0033291721002609.
26. Clifford S, Wendell DG. How disgust influences health purity attitudes. *Polit Behav*. 2016;38(1):155–178.
27. Kerr JR, Schneider CR, Recchia G, et al. Correlates of intended COVID-19 vaccine acceptance across time and countries: Results from a series of cross-sectional surveys. *BMJ Open*. 2021;11, e048025. https://doi.org/10.1136/bmjopen-2020-048025.
28. Motta M, Callaghan T, Sylvester S, Lunz TK. Identifying the prevalence, correlates, and policy consequences of anti-vaccine social identity. *Polit, Groups, Ident*. 2021;1–15.
29. Huddy L. Group identity and political cohesion. In: Sears DO, Huddy L, Jervis R, eds. *Oxford Handbook of Political Psychology*. Oxford University Press; 2003:511–558.
30. Van Prooijen JW, Douglas KM. Belief in conspiracy theories: Basic principles of an emerging research domain. *Eur J Soc Psychol*. 2018;48(7):897–908.
31. Paterson P, Chantler T, Larson HJ. Reasons for non-vaccination: Parental vaccine hesitancy and the childhood influenza vaccination school pilot programme in England. *Vaccine*. 2018;36(36):5397–5401.
32. Freimuth VS, Jamison AM, An J, Hancock GR, Quinn SC. Determinants of trust in the flu vaccine for African Americans and whites. *Soc Sci Med*. 2017;193:70–79.

33. Bajaj SS, Stanford FC. Beyond Tuskegee – Vaccine distrust and everyday racism. *N Engl J Med.* 2021;384(5): e12.
34. Kennedy J. Populist politics and vaccine hesitancy in Western Europe: An analysis of national-level data. *Eur J Public Health.* 2019;29(3):512–516.
35. Hornsey MJ, Finlayson M, Chatwood G, Begeny CT. Donald Trump and vaccination: The effect of political identity, conspiracist ideation and presidential tweets on vaccine hesitancy. *J Exp Soc Psychol.* 2020;88, 103947.
36. Motta M. Republicans, not democrats, are more likely to endorse anti-vaccine misinformation. *Am Politics Res.* 2021;49(5):428–438.
37. Goldberg ZJ, Richey S. Anti-vaccination beliefs and unrelated conspiracy theories. *World Aff.* 2020;183(2):105–124.
38. Motta M, Stecuła D. Quantifying the effect of Wakefield et al. (1998) on skepticism about MMR vaccine safety in the U.S. *PLoS One.* 2021;16(8):e0256395. https://doi.org/10.1371/journal.pone.0256395.
39. Allington D, McAndrew S, Moxham-Hall V, Duffy B. Coronavirus conspiracy suspicions, general vaccine attitudes, trust and coronavirus information source as predictors of vaccine hesitancy among UK residents during the COVID-19 pandemic. *Psychol Med.* 2021;1–12. https://doi.org/10.1017/S0033291721001434.
40. Farhart C, Chen P. Racialized pandemic: The effect of racial attitudes on COVID-19 conspiracy theory beliefs. *Front in Polit Sci.* 2021.
41. Hartman TK, Marshall M, Stocks TV, et al. Different conspiracy theories have different psychological and social determinants: Comparison of three theories about the origins of the COVID-19 virus in a representative sample of the UK population. *Front in Polit Sci.* 2021;3. https://doi.org/10.3389/fpos.2021.642510.
42. Jennings W, Stoker G, Bunting H, et al. Lack of trust, conspiracy beliefs, and social media use predict COVID-19 vaccine hesitancy. *Vaccine.* 2021;9(6):593. https://doi.org/10.3390/vaccines9060593.
43. Ognyanova K, Lazer D, Baum M, et al. *The COVID States Project #60: COVID-19 Vaccine Misinformation: From Uncertainty to Resistance*; 2021. https://doi.org/10.31219/osf.io/xtjad.
44. Palm R, Bolsen T, Kingsland JT. The effect of frames on COVID-19 vaccine resistance. *Front in Polit Sci.* 2021;3. https://doi.org/10.3389/fpos.2021.661257.
45. Vitriol JA, Marsh JK. A pandemic of misbelief: How beliefs promote or undermine COVID-19 mitigation. *Front Polit Sci.* 2021;3. https://doi.org/10.3389/fpos.2021.648082.
46. Allington D, Duffy B, Wessely S, Dhavan N, Rubin J. Health-protective behaviour, social media usage and conspiracy belief during the COVID-19 public health emergency. *Psychol Med.* 2020;51(10):1–7. https://doi.org/10.1017/S003329172000224X.
47. Motta M, Callaghan T, Sylvester S. Knowing less but presuming more: Dunning-Kruger effects and the endorsement of anti-vaccine policy attitudes. *Soc Sci Med.* 2018;211:274–281. https://doi.org/10.1016/j.socscimed.2018.06.032.
48. Bode L, Vraga EK. See something, say something: Correction of global health misinformation on social media. *Health Commun.* 2017;33(9):1131–1140.
49. Lazer DMJ, Baum MA, Benkler Y, et al. The science of fake news. *Science.* 2018;359 (6380):1094–1096.
50. Keelan J, Pavri-Garcia V, Tomlinson G, Kumanan W. YouTube as a source of information on immunization: A content analysis. *JAMA.* 2007;298(21):2482–2484. https://doi.org/10.1001/jama.298.21.2482.
51. Dunn AG, Surian D, Leask J, Dey A, Mandl KD, Coiera E. Mapping information exposure on social media to explain differences in HPV vaccine coverage in the United States. *Vaccine.* 2017;35(23):3033–3040.

52. Enders AM, Uscinski JE, Seelig MI, et al. The relationship between social media use and beliefs in conspiracy theories and misinformation. *Polit Behav*. 2021. https://doi.org/10.1007/s11109-021-09734-6.
53. Dror AA, Eisenbach N, Taiber S, et al. Vaccine hesitancy: The next challenge in the fight against COVID-19. *Eur J Epidemiol*. 2020;35(8):775–779.
54. Fadda M, Albanese E, Suggs LS. When a COVID-19 vaccine is ready, will we all be ready for it?. *Int J Public Health*. 2020;65:711–712.
55. Green J, Edgerton J, Naftel D, Shoub K, Cranmer SJ. Elusive consensus: Polarization in elite communication on the COVID-19 pandemic. *Sci Adv*. 2020;6(28): [eabc2717].
56. Kreps S, Prasad S, Brownstein JS, et al. Factors associated with US adults' likelihood of accepting COVID-19 vaccination. *JAMA Netw Open*. 2020;3(10): [e2025594.].
57. Kates J, Tolbert J, Orgera K. *The red/blue divide in COVID-19 vaccination rates is growing*. Kaiser Family Foundation; 2021. Accessed September 9, 2021. https://www.kff.org/policy-watch/the-red-blue-divide-in-covid-19-vaccination-rates-is-growing/.
58. Miller JM. Psychological, political, and situational factors combine to boost COVID-19 conspiracy theory beliefs. *Can J Polit Sci*. 2020;53:327–334. https://doi.org/10.1017/S000842392000058X.
59. Cassese EC, Farhart CE, Miller JM. Gender differences in COVID-19 conspiracy theory beliefs. *Polit Gend*. 2020;16(4):1009–1018.
60. Lunz TK. *A Case of Misunderstood Identity: The Role of Rural Identity in Contemporary American Mass Politics. Dissertation*. Twin Cities: University of Minnesota; 2021.
61. Motta M. The dynamics and political implications of anti-intellectualism in the United States. *Am Politics Res*. 2018;46(3):465–498.
62. Merkley E, Loewen PJ. Anti-intellectualism and the mass public's response to the COVID-19 pandemic. *Nat Hum Behav*. 2021;5:706–715. https://doi.org/10.1038/s41562-021-01112-w.
63. Berinsky AJ, Huber GA, Lenz GS. Evaluating online labor markets for experimental research: Amazon.com's mechanical Turk. *Polit Anal*. 2012;20(3):351–368.
64. Mason W, Suri S. Conducting behavioral research on Amazon's mechanical Turk. *Behav Res Methods*. 2012;44(4):1–23.
65. Funk C, Tyson A. Growing share of Americans say they plan to get a COVID-19 vaccine – Or already have. *Pew Research Center MARCH*. 2021;5. https://www.pewresearch.org/science/2021/03/05/growing-share-of-americans-say-they-plan-to-get-a-covid-19-vaccine-or-already-have/.
66. KFF. *COVID-19 Vaccine Monitor. Kaiser Family Foundation*; 2021. Accessed September 16, 2021 https://www.kff.org/coronavirus-covid-19/dashboard/kff-covid-19-vaccine-monitor-dashboard/#intentions.
67. Betsch C, Schmid P, Heinemeier D, Korn L, Holtmann C, Böhm R. Beyond confidence: Development of a measure assessing the 5C psychological antecedents of vaccination. *PLoS One*. 2018;13(12): e0208601. https://doi.org/10.1371/journal.pone.0208601.
68. Uscinski JE. *Conspiracy Theories: A Primer*. Rowman & Littlefield; 2020.
69. Uscinski JE, Parent JM. *American Conspiracy Theories*. Oxford University Press; 2014.
70. Uscinski JE, Klofstad C, Atkinson MD. What drives conspiratorial beliefs? The role of informational cues and predispositions. *Polit Res Q*. 2016;69(1):57–71.
71. Barker DC, Detamble R, Morgan M. Intellectualism, anti-intellectualism, and epistemic hubris in red and blue America. *Am Polit Sci Rev*. 2021;1–16. https://doi.org/10.1017/S0003055421000988.
72. Oliver JE, Rahn WM. Rise of the "Trumpenvolk": Populism in the 2016 election. *Ann Am Acad Pol Soc Sci*. 2016;667(1):189–206.
73. Farhart CE. *Look Who Is Disaffected Now: Political Causes and Consequences of Learned Helplessness in the U.S.* Dissertation. Twin Cities: University of Minnesota; 2017.

74. Farhart CE, Miller JM, Saunders KL. Conspiracy stress or relief? Learned helplessness and conspiratorial thinking. In: Suhay E, Barker D, eds. *The Politics of Truth in Polarized America*. Oxford: Oxford University Press; 2021.
75. Torcal M, Montero JR, eds. *Political Disaffection in Contemporary Democracies: Social Capital, Institutions, and Politics*. Routledge; 2006.
76. Rosenthal A, Motta M, Farhart CE. *Beyond Tuskegee, to Middlesboro: How Perspectives of Policing Shape Vaccine Attitudes for Black Americans*; 2021. https://doi.org/10.31235/osf.io/wjq4a.
77. Craig SC, Niemi RG, Silver GE. Political efficacy and trust: A report on the NES pilot study items. *Polit Behav*. 1990;12(3):289–314.
78. Peterson C, Maier SF, Seligman MEP. *Learned Helplessness: A Theory for the Age of Personal Control*. Oxford University Press; 1993.
79. Quinless FW, McDermott Nelson MA. Development of a measure of learned helplessness. *Nurs Res*. 1988;37(1):11–15.
80. Di Palma G. Disaffection and participation in Western democracies: The role of political oppositions. *J Theor Polit*. 1969;31(4):984–1010.
81. Delli Carpini MX, Keeter S. *What Americans Know about Politics and why it Matters*. Yale University Press; 1996.
82. Vitriol JA, Tagar MR, Federico CM, Sawicki V. Ideological uncertainty and investment of the self in politics. *J Exp Soc Psychol*. 2019;82:85–97.
83. Miller JM, Saunders KL, Farhart CE. Conspiracy endorsement as motivated reasoning: The moderating roles of political knowledge and trust. *Am J Polit Sci*. 2016;60 (4):824–844.
84. Hotez PJ, Cooney RE, Benjamin RM, et al. Announcing the lancet commission on vaccine refusal, acceptance, and demand in the USA. *The Lancet*. 2021;397 (10280):1165–1167. https://doi.org/10.1016/S0140-6736(21)00372-X.
85. Stecula DA, Pickup M. How populism and conservative media fuel conspiracy beliefs about COVID-19 and what it means for COVID-19 behaviors. *Res Polit*. 2021;8 (1):1–9. https://doi.org/10.1177/2053168021993979.
86. Romer D, Jamieson KH. Patterns of media use, strength of belief in COVID-19 conspiracy theories, and the prevention of COVID-19 from March to July 2020 in the United States: Survey study. *J Med Internet Res*. 2021;23(4):e25215. Published 2021 Apr 27. https://doi.org/10.2196/25215.
87. Romer D, Jamieson KH. Conspiracy theories as barriers to controlling the spread of COVID-19 in the U.S. *Soc Sci Med*. 2020;263, 113356. https://doi.org/10.1016/j.socscimed.2020.113356.
88. Pennycook G, McPhetres J, Bago B, Rand DG. *Beliefs about COVID-19 in Canada, the U.K., and the U.S.A.: A Novel Test of Political Polarization and Motivated Reasoning*; 2021. https://doi.org/10.31234/osf.io/zhjkp.
89. Imhoff R, Lamberty P. A bioweapon or a hoax? The link between distinct conspiracy beliefs about the coronavirus disease (COVID-19) outbreak and pandemic behavior. *Soc Psychol Personal Sci*. 2020;11(8):1110–1118. https://doi.org/10.1177/1948550620934692.
90. Marinthe G, Brown G, Delouvée S, Jolley D. Looking out for myself: Exploring the relationship between conspiracy mentality, perceived personal risk, and COVID-19 prevention measures. *Br J Health Psychol*. 2020;25(4):957–980.
91. Lyons B, Merola V, Reifler J. Not just asking questions: Effects of implicit and explicit conspiracy information about vaccines and genetic modification. *Health Commun*. 2019;34(14):1741–1750.
92. Carey JM, Chi V, Flynn DJ, Nyhan B, Zeitzoff T. The effects of corrective information about disease epidemics and outbreaks: Evidence from zika and yellow fever in Brazil. *Sci Adv*. 2020;6(5): eaaw7449. https://doi.org/10.1126/sciadv.aaw7449.

93. Compton J, van der Linden S, Cook J, Basol M. Inoculation theory in the post-truth era: Extant findings and new frontiers for contested science, misinformation, and conspiracy theories. *Soc Personal Psychol Compass.* 2021;15, e12602. https://doi.org/10.1111/spc3.12602.
94. Lewandowsky S, van der Linden S. Countering misinformation and fake news through inoculation and Prebunking. *Eur Rev Soc Psychol.* 2021. https://doi.org/10.1080/10463283.2021.1876983.
95. Islam MS, Kamal A-HM, Kabir A, et al. COVID-19 vaccine rumors and conspiracy theories: The need for cognitive inoculation against misinformation to improve vaccine adherence. *PLoS One.* 2021;16(5): e0251605. https://doi.org/10.1371/journal.pone.0251605.
96. Pennycook G, McPhetres J, Zhang Y, Lu JG, Rand DG. Fighting COVID-19 misinformation on social media: Experimental evidence for a scalable accuracy-nudge intervention. *Psychol Sci.* 2020;31(7):770–780.
97. Biddlestone M, Green R, Douglas KM. Cultural orientation, powerlessness, belief in conspiracy theories, and intentions to reduce the spread of COVID-19. *Br J Soc Psychol.* 2020;59:663–673.
98. Van Bavel JJ, Baicker K, Boggio PS, et al. Using social and behavioural science to support COVID-19 pandemic response. *Nat Hum Behav.* 2020;4:460–471.
99. Jolley D, Douglas KM. Prevention is better than cure: Addressing anti-vaccine conspiracy theories. *J Appl Soc Psychol.* 2017;47(8):459–469.
100. Roozenbeek J, van der Linden S, Nygren T. *Prebunking Interventions Based on "Inoculation" Theory Can Reduce Susceptibility to Misinformation across Cultures*; 2020. The Harvard Kennedy School (HKS) Misinformation Review https://doi.org/10.37016//mr-2020-008.

CHAPTER EIGHT

Communicating CRISPR: Challenges and opportunities in engaging the public

Elizabeth S. Baik, Abraham Koshy, and Bruce W. Hardy*
Klein College of Media and Communication, Temple University, Philadelphia, PA, United States
*Corresponding author: e-mail address: bruce.hardy@temple.edu

Contents

1. Introduction 172
2. The challenge from the speed of advancement 173
3. What CRISPR means to different publics: The challenge of lack of quality data 174
4. Engaging diverse and underrepresented publics: Shared knowledge production and decision-making 178
5. Public engagement of CRISPR in the age of politicization 181
6. Ways to overcome identity protective cognition 183
   6.1 Priming different identities 183
   6.2 Abstract versus concrete cognition 184
   6.3 The LIVA model for science communication 186
7. Conclusion 187
References 188

Abstract

CRISPR technologies are advancing at a dizzying pace, and emerging cultural, sociopolitical, ethical, and legal implications continue to pose new challenges for public engagement. Recent calls for public engagement and dialogue on CRISPR applications stress the importance of nuanced thinking and responsible communication. In this chapter, we review public opinion research and find that a comprehensive and clear picture of global views on CRISPR is missing but is necessary to build the foundation for effective public engagement programs. We recommend community-based-participatory research as an inclusive and effective framework for shared knowledge production and decision-making practices for scientific experts and science communicators to engage in genuine and meaningful dialogue with community members in making informed consideration for important value-laden decisions. In response to the politicization of science, this chapter offers strategic communication techniques that can help those facilitating public engagement of CRISPR-based technologies keep cognitive biases, such as identity protective cognition, motivated reasoning, and confirmation bias, at bay.

Progress in Molecular Biology and Translational Science, Volume 188
ISSN 1877-1173
https://doi.org/10.1016/bs.pmbts.2021.11.004

1. Introduction

In September 2021, the US Food and Drug Administration approved a CRISPR-based gene editing therapy for clinical trials in human patients with HIV infections.[1] In collaboration with Excision BioTherapeutics,[1] academic researchers at Temple University developed and tested this groundbreaking CRISPR-based system that had reduced proviral DNA in SIV-infected macaques.[2] Should the clinical trials prove successful, this CRISPR-based therapy would cure HIV infection in "a single intravenous infusion."[3]

Earlier in the same year, in April 2021, researchers at the University of California San Francisco and the Whitehead Institute advanced CRISPR technology to reach beyond the human genome and into the epigenome.[4] These new CRISPR tools called "CRISPRoff" and "CRISPRon" can turn off and on any gene in the human genome without editing the genetic code. The researchers demonstrated that CRISPRoff can efficiently silence genes with high specificity. "Once a gene is switched off, it remains inert in the cell's descendants for hundreds of generations, unless it is switched back on with a complimentary" CRISPRon tool.[5] This on-off switch for gene editing can command gene expression without altering the underlying DNA sequence, unlike traditional CRISPR-Cas9 sequencing that makes permanent modifications to the genetic code by cutting and editing DNA strands.

These are just two recent examples of CRISPR-based innovations that occurred in the last year. CRISPR (an acronym for Clustered Regularly Interspaced Short Palindromic Repeats) technologies are precise gene-editing tools that are advancing at a dizzying pace, and fast-emerging cultural, sociopolitical, ethical, and legal implications continue to pose new challenges for public engagement. These challenges involve public concerns about not only existing CRISPR technologies, but also unforeseen ramifications of yet to come CRISPR advancements in plant, animal, and human biology. Communicating science is a messy business, as has become apparent throughout the COVID-19 pandemic, during which the politicization of science intensified the spread of misinformation[6] and conspiracy theories,[7,8] an erosion of deference to science and public health officials,[9] and frustration with simple health behaviors, such as wearing a mask, in the United States.[10] Take for example the myths and misinformation currently fostering COVID-19 vaccine hesitancy; because the Pfizer and Moderna mRNA

COVID vaccines were built on scientific breakthroughs ushered by CRISPR technologies, false claims that the vaccines would genetically modify people who receive them have spread widely on social media.[11] As such, future public dialogue on new CRISPR technologies could become polluted.

Recent calls for public engagement and dialogue on CRISPR applications stress the importance of nuanced thinking and responsible communication given that CRISPR technologies are "post-normal" science,[12,13] where the "facts are uncertain, values in dispute, stakes high and decisions urgent."[14] Post-normal science requires much more than just technical expertise to understand and address the intricacies of the sociopolitical and ethical complexities around scientific advancements like CRISPR. In reference to CRISPR as post-normal science, Scheufele and colleagues[13] argued, "Decision stakes are high, and margins of error are thin, especially once we cross the bright red line of editing the human germline and begin making edits heritable." Therefore, voices and perspectives from multiple and diverse publics need to be considered when assessing the benefits and risks of new CRISPR-based advancements.

2. The challenge from the speed of advancement

As an indicator of the speed at which CRISPR technologies are rapidly advancing, consider the fact that the US National Institute for Health (NIH) increased its funding for CRISPR-related research from $5 million in fiscal year 2011 to $1.1 billion in fiscal year 2018.[15] While CRISPR technologies have been mostly developed in academic institutions, there is a trend of scientists based in these academic institutions creating "spin out CRISPR biotechnology companies" to generate capital and "comprehensive intellectual property portfolios to rapidly commercialize CRISPR products."[16] The genome editing global market size is projected to reach $13.83 billion by 2028 at a compound annual growth rate of 14.53%.[17] As a recent example of the commercialization of CRISPR technologies,[16] a CRISPR startup called Prime Medicine raised $325 million, mostly from repeat investors, to develop a "Crispr Swiss Army knife."[18] The infusion of governmental funding, venture capital, and other investments into CRISPR technologies will fuel the rapid speed of advancement.

Barriers to meaningful public engagement around CRISPR technologies are only heightened by the speed of advancement. "It has created a bottleneck problem for societies trying to address the host of ethical, societal, and regulatory questions that arise almost simultaneously."[13] What this means is that those facilitating public engagement on CRISPR-based technologies are constantly lagging behind. Successful dialogue with the full range of stakeholders becomes near impossible because it is too difficult to keep up with the pace of advancements in CRISPR technologies. "Traditionally, ethical reflection has come only after research and development, and the application of innovations was slow enough that this discussion could be far advanced before their full impact has been felt."[19] Ethical reflections can no longer remain on the back burner while conducting research. Effective public engagement and considerations over the cultural, sociopolitical, ethical, and legal implications should happen at all stages of development and research of CRISPR-based technologies. Every step of the process should be informed by reflections about societal impacts of the new advancement.

3. What CRISPR means to different publics: The challenge of lack of quality data

One of the first things we need to understand is what CRISPR-based technology means to different publics. Currently, we have very little insight into public views of CRISPR technologies. A recent review[20] of public opinion polls on human gene editing reveals only a handful of quality and methodologically sound surveys have been conducted since the development of CRISPR. Out of the 10 surveys identified in this review, six of those were conducted in the United States, indicative of the lack of geographical and cultural representation and therefore limitations of knowledge produced. Not included in the review[20] are studies that rely on non-scientific sampling procedures (e.g., convenience sampling) as well as a handful of large-scale, but not representative, global studies conducted on people's views on CRISPR.[21,22] Some of these studies that rely on non-probability sampling, such as that of McCaughey and colleagues,[22] rely on recruiting respondents through social media, leaving the generalizations of the results limited because it is difficult to understand who the participants represent. The current state of knowledge of public views and understanding of CRISPR technology is therefore inadequate.

A comprehensive and clear picture of global views on CRISPR is missing and necessary to build the foundation for effective public engagement programs. A 2015 US National Academies of Sciences, Engineering, and

Medicine (NASEM) international summit on regulating CRISPR technologies echoed the need for understanding global public sentiment. The NASEM released a statement from the summit saying, "It would be irresponsible to proceed with any clinical use of germline editing unless and until (i) the relevant safety and efficacy issues have been resolved, based on appropriate understanding and balancing of risks, potential benefits, and alternatives, and (ii) there is broad societal consensus about the appropriateness of the proposed application."[23]

Notwithstanding the current shortage of representative and quality data, existing studies on public views of CRISPR signal a trend: People are more supportive of therapeutic CRISPR applications than of enhancement applications. In a nationally represented survey of 1600 US adults, Scheufele and his colleagues[24] found that 59% of respondents supported human gene editing to treat a medical condition or to restore health, compared to only 33% of respondents who supported the same to improve human abilities. Similar findings come from the Pew Research Center, where a 2018 survey[25] found that 7 in 10 Americans (72%) believe that changing a baby's genes to "treat a serious disease/condition the baby would have at birth" is an appropriate application of CRISPR technologies as opposed to 27% who reported that doing so would "take the medical technology too far." When asked about human enhancement by CRISPR-based technologies, four-fifths (80%) of those surveyed said that making a baby more intelligent would "take the medical technology too far."[25] Surveying people from Austria, Denmark, Germany, Hungary, Iceland, Italy, the Netherlands, Portugal, Spain, UK and the United States, Gaskell et al.[26] also found consistent support for therapeutic CRISPR applications over enhancement applications. Wang et al.[27] found similar attitudes toward CRISPR in China as did Critchley et al.[28] in Australia. This difference in support of therapeutic and enhancement application is also found in large-scale, global studies, albeit their non-representative samples.[21,22] All studies that were included in a systematic review by Delhove et al.[29] also found that people held more positive attitudes toward CRISPR-based therapy or treatment than toward enhancement.

There is some evidence—and it makes sense—that support for CRISPR-based technology varies across different sociodemographics.[29] But given the limited amount of quality research that has been conducted on the public views of CRISPR, studies looking at demographic differences yield mixed results. Delhove et al.[29] point to a handful of studies that suggest men may be more approving of CRISPR than females may be, while other studies found no significant differences nor relationships. The same systematic review[29] found that education was a strong positive predictor of approval of CRISPR technologies in the majority of reviewed studies, while age was a less certain predictor.

Religiosity has also been examined and studies have found it to be a negative predictor of support for CRISPR. Both Scheufele et al.[24] and Pew Research Center[25] report this negative link between religious status and support for CRISPR-based applications in the United States. But we caution against generalizing this negative relationship to communities outside of the United States. The relationship between religiosity and science in the United States is strained with those who are religious holding less positive attitudes toward science than those who are not.[30,31] Importantly this negative relationship between religiosity and science varies from country to country. McPhetres, Jong, and Zuckerman[30] examined data from over 60 countries and found positive, negative, and null relationships between religiosity and science. One recent study found that the relationship between religiosity and a lack of confidence in science may have to do with increasing levels of those who identify as religiously unaffiliated in a given country. Those who are religious in countries where they are a demographic minority to the religiously unaffiliated are far more skeptical of science than the religious who are the majority in their country.[32] This reinforces the argument that those who wish to facilitate public engagement on CRISPR cannot extrapolate knowledge from US studies to non-US populations. Overall, quality public opinion research on a global scale is desperately needed.

CRISPR advancements also highlight the need for genetic literacy[33,34] to facilitate educated deliberation and informed decision-making. While there are a handful of public opinion polls about support, there is surprisingly little quality research on public knowledge of CRISPR technologies. As of this writing, we could not find any scientific studies using probability-based samples to gauge public knowledge on CRISPR. This is a major shortcoming to our understanding of what CRISPR means to different publics. One study using non-scientific sampling, however, analyzed an international sample of 3988 open-ended responses, and through topic modeling revealed that respondents who were skeptical of CRISPR applications overwhelmingly talked about a need for "better understanding" of the technology.[35] Simply put, we have almost no insight into public understanding of CRISPR technologies.

An important area for science communications researchers to tackle in the quest for understanding what CRISPR means to different publics is examining how CRISPR is being reported and portrayed in media in the community where public engagement is to be facilitated. Like the lack of research on public views and public knowledge of CRISPR, very little research has been published on media coverage of it. A recent content analysis of CRISPR in the North American popular press did find that most of the coverage (83% of articles in the sample) discussed CRISPR in terms of human health and only 26.8% of the articles touched upon ethical, legal, and/or social issues.[36] Understanding what aspects of CRISPR-based technologies are being made salient in media can tell those engaging in public outreach which aspects of CRISPR are salient to the public. Decades of research on influence of news media in agenda-setting and priming[37,38] inform us that the attributes made salient about an issue in news coverage will influence the public's judgments on policy. So, if CRISPR-based technologies are being mostly reported in terms of human health in news media, public engagement and discussion on CRISPR—even if this engagement is related to animal or plant biotechnology—will be influenced by the salience of CRISPR in relation to human health caused by media coverage of it. Knowledge of how CRISPR technology is being discussed in the media environment is important for understanding public evaluation of the technology.

Take for example the news coverage that the Chinese gene-edited babies scandal commanded in 2018; on November 25, Chinese scientist He Jiankui announced that twin babies were born in China with genetically edited CCR5 genes to protect them against HIV. The announcement provoked widespread criticism[39] and news coverage. "Such a scientific crisis can be a significant threat to the reputation of relevant scientists and even the whole scientific community which can reshape the way various social actors perceive and discuss gene editing."[40] One study found that the scandal did increase awareness of genome editing but decreased acceptance.[41] A content analysis[40] of 2074 posts on the Chinese social media site Weibo indicated that "laypeople" did have more discussions on gene editing after the scandal, but that emotions like anger, shame, and disgust also significantly increased, as did metaphors like "mad scientist" and "Pandora Box." Calabrese et al.'s analysis[42] of tweets before, during, and after the story about He Jiankui broke in the news revealed a sharp negative turn in the sentiment of tweets following the event. This echoes findings by Müller et al.,[43] whose analysis of 1.3 billion tweets since 2013 revealed a slow negative trend in sentiment expressed toward CRISPR in tweets, with negative peaks in sentiment around highly publicized news stories like the one about He Jiankui.

Given the few studies on public perception of CRISPR, we can only piece together a very rough sketch of what CRISPR means to different publics with much of this knowledge being generated by studies relying on US samples.[20] There are many publics that will be affected by CRISPR-based technologies and we currently have absolutely no insight into their perceptions or understanding of what it is. As of this writing, there have been no studies conducted anywhere in South America or Africa published or translated in English.[20] There is an urgent need for quality data from representative samples from these parts of the world to inform any type of public engagement on CRISPR applications. It is surprising just how little quality research there has been done.

4. Engaging diverse and underrepresented publics: Shared knowledge production and decision-making

CRISPR-based technologies are often heralded as a cure for conditions that disproportionately impact marginalized communities. For example, genetically engineered mosquitoes are developed to potentially suppress populations of mosquitoes that carry the malaria virus.[44] Additionally, applications of prenatal genetic technologies aim to address genes that can cause deafness and Down syndrome in babies.[45,46] Without effective public engagement, however, the intense speed at which CRISPR and its applications have developed may leave behind the very people and communities that the scientific advancements ostensibly benefit.[13] Communicating CRISPR must engage underrepresented communities with intention,[47] and thereby widen and deepen the representation of voices in developing technologies and strategies.[13]

Meaningful communication is built upon collaborative partnership and shared decision-making processes. In the case of gene research for malaria control, the NASEM has urged the field to shift away from a knowledge-deficit approach, which views communication as a top-down transmission of information.[44] Instead, effective communication of CRISPR-based technologies involves *knowledge engagement*, whereby stakeholders at all levels have agency and ownership in knowledge production. As Africa carries the vast majority of the global malaria burden,[48] applications of gene technology for malaria control necessitate expertise in local epidemiology and entomology.[44] Scientists from Tanzania and Kenya have called upon African research institutions to take the lead for "malaria-afflicted nations to regain their roles as stakeholders, decisionmakers, and eventual owners of this technology."[49] Expert publics must equitably involve local community members and experts as co-creators and owners of knowledge. Such knowledge engagement is not merely a moral obligation nor an expedient avenue to local acceptance, but rather a way to generate diverse perspectives and knowledge that contribute to better technologies and solutions that meet communities where they are.

Within health research, academics and practitioners have delineated processes of knowledge engagement with the conceptual model of community-based participatory research (CBPR). As a collaborative research approach, CBPR's central ethos is to redefine relations among academics, policy makers, and community members by redistributing power and resources.[50] In doing so, CBPR creates "mutual trust, culturally centered research methods, sustainable interventions, and community capacity."[50] CBPR's core principles include (1) centering and understanding the strengths of a community for which research is conducted, (2) building equitable partnerships based on long-term commitment, (3) developing and applying interventions that are mutually beneficial for every stakeholder involved, and (4) capacity building for sustainable change.[51] These processes of partnership, research, and change are cyclical and iterative as complex systems of a community continue to evolve.

As a research orientation, CBPR is not limited to specific qualitative and quantitative research methods.[50] Just as the research question or problem should determine the appropriate research method, it is only appropriate for CRISPR research that seeks to benefit a given community to utilize a research orientation that involves community members in the research process. Scientists, academics, and practitioners involved in CRISPR research must consider the role and function of CBPR in increasing public engagement and generating community-based solutions. CRISPR research that seeks large-scale solutions with far-reaching—and often irreversible—effects must center marginalized communities in its knowledge production and decision-making process.

If sharing resources and power in knowledge production raises practical and epistemological concerns, then researchers may begin by acknowledging and reflecting on potentially paternalistic attitudes. At a minimum, shared knowledge production and decision-making practices require scientific experts to engage in genuine and meaningful dialogue so that the community members can make informed consideration for important value-laden decisions. CRISPR may offer possibilities of corrections, modifications, or silencing of "undesirable" genes in humans. Yet, such notions that certain genetic expressions are undesirable persistently center dominant cultural views. These positions of scientific paternalism will continue to run into the deeply subjective nature of value assessments rooted in various identities and belongings of individuals. Much like medical paternalism, by which health professionals disregard the patient's individual values and preferences on the basis of their own medical expertise, scientific paternalism permeates and influences practices in knowledge production and application. Both overt and unacknowledged paternalistic notions that one's expertise and therefore perspectives take precedence over another individual's values and preferences result in such decisions as to which "undesirable" genes should be "corrected" or "silenced."

These attitudes are apparent in the language used in describing CRISPR-based applications and interventions. Writing about the mutated gene TMC1 that causes deafness, Yeh and colleagues[45] report CRISPR as having "the potential to directly repair point mutations and provide therapeutic restoration of gene function." Yet, deaf communities may reject such notions that their genes require repair and restoration.[52] Deaf communities across the globe are thriving cultural and linguistic groups that contribute to biocultural diversity. Baumann and Murray[53] argue that a genetic mutation that causes deafness should not be narrowly seen as a disadvantage, as such genetic mutation also involves unexpected advantages. For example, the Cx26 gene that causes hereditary deafness is also "responsible for increased protection against infections from bacteria as well as accelerated healing of wounds."[53] This deaf gain perspective[53] reframes deafness as a biological *difference* rather than a *deficiency*.[52] It is crucial to consider the necessity and benefits of such biocultural differences and diversity made only possible through varied genetic mutations and expressions.

Similar to deafness, Down syndrome is considered a disability in dominant cultures, and therefore a target for CRISPR application. However, Down syndrome may be an integral factor of one's identity that enriches the lives of individuals in and around these communities. A survey of parents of people with Down syndrome revealed that a strong majority of the families enjoyed a high quality of life (90.7%) and considered Down syndrome a positive force in their lives (71.6).[46] Parents were more ambivalent, however, when prompted to consider a prenatal intervention that would silence the expression of the extra chromosome that causes Down syndrome, with 48.1% opposing the intervention.[46] When communicating CRISPR-based solutions to relevant publics, it is important to consider which integral aspects of one's identity and those of loved ones is the intervention threatening to erase. As for parents of people with Down syndrome, the threat is twofold. The intervention would not only modify the person with Down syndrome, thereby raising anxieties and doubt of a parent's self-image to love a child unconditionally, but also possibly destabilize existing family dynamics, relations, and identities established upon years of deeply involved caregiving practices. CRISPR may be a simple solution for the researching scientist, but poses complex challenges for potential recipients.

This is not to make sweeping statements about marginalized groups, but rather to urge researchers to consult and involve underrepresented perspectives in CRISPR research and applications. For example, the Environmental Protection Authority (EPA) of New Zealand had appointed a Māori advisory committee named Ngā Kaihautū Tikanga Taiao that facilitates and ensures the inclusion of indigenous perspectives in the EPA's approval of gene editing.[54] Contrary to conventional belief and previous literature that stereotype indigenous people as opposing gene editing, Māori informants conveyed their belief that CRISPR technologies may complement their cultural values, such as *whakapapa* (genealogy), *mana* (power), *mauri* (life essence), and *kaitiakitanga* (guardianship).[54] Such findings confirm the "need for more research to understand how the practices, beliefs, previous knowledge and experiences, and values of marginalized individuals influence how they engage with, perceive, and understand gene editing."[47]

5. Public engagement of CRISPR in the age of politicization

As the title of this volume (*Molecular Biology and Clinical Medicine in the Age of Politicization*) indicates, scientific debates have become politicized.[55,56] There is a growing body of work that shows that, particularly in the United States, political ideology reliably predicts beliefs about politicized scientific issues, such as climate change,[57,58] evolution, the Big Bang Theory,[59] and the overall credibility of scientists.[60] As we discussed earlier, the divergence of belief about COVID-19 by ideology in the United States[8] is endemic to the politicization of science. Although there are studies suggesting that liberals are equally as likely as conservatives to reject scientific evidence that is incongruent with their beliefs,[61] most research in this area has generally pointed to an overall ideological divide with liberals being more likely than conservatives to embrace science. A recent study found that liberals were much more likely than conservatives to hold positive evaluations of science and scientists without any reference to a specific scientific issue, and that this difference was conditioned on level of interest in politics and news use.[62] Such embracing of science by those on the left has become almost a type of political ownership of science: "the Democrats found that science was a good thing for them, just like historically strong support for the military was good for the Republicans."[63] On the other hand, because science is often in conflict with cultural values and core beliefs of political conservatism, such as the laissez-faire economic policy[64] and a belief in God,[65] those on the political right in the United States often reject science to protect these core beliefs and values.

In the United States, the COVID-19 pandemic exacerbated political divisions around science and public health. In inoculating the country against the coronavirus, vaccine mandates have faced considerable backlash from those who hold a conservative ideology because the mandates are seen as imposing on personal freedom and liberty—a core of the American ideal. An October 2021 poll of US adults found that when asked about a mandate for childhood COVID-19 vaccinations, "79% of Democrats are in favor and 9% are opposed, compared to 25% and 63% for Republicans," respectively.[66] According to Kaiser Health News, 26 states had passed laws during the pandemic that significantly weakened the power of state and local public health officials.[67]

Unfortunately, there is little reason to believe that if CRISPR-based advancements reach a national dialogue in the United States that it will be immune to politicization. When an issue becomes politicized, information and messages about the issue gets interpreted through identity protective lens,[68] which fosters confirmation bias and motivated reasoning.[69] The identity protective cognition thesis states that people's beliefs and attitudes about a polarized issue fall in line with the beliefs of their affinity group. Conversely, when a person is presented with information that goes against the beliefs of their affinity group, they will reject the information. "Because of such identity protective cognition, when an issue is politically polarized, it can be unproductive or even counter-productive to provide new information that contradicts an individuals' pre-existing viewpoint."[70] This, of course, will introduce challenges for public engagement.

Interestingly, CRISPR-based technologies have not yet reached a level of national dialogue that provokes politicization of the science behind it. Surprisingly, the announcement of a possible CRISPR-based cure for HIV by scientists at Temple University and Excision BioTherapeutics, mentioned in the opening of this chapter, only garnered a small amount of news coverage—mostly from local and medical/science news outlets. So, while there is a lot of discussion in the scientific and academic community, CRISPR-based technologies have not entered a national debate in the United States where it becomes infected with political in-group/out-group division or affective partisanship. Yet, it is not hard to imagine that this will ultimately happen to public opinion and debate around CRISPR-technologies. Once it does, there are strategic communication techniques that can help those facilitating public engagement to keep identity protective cognition, motivated reasoning, and confirmation bias at bay.

6. Ways to overcome identity protective cognition

6.1 Priming different identities

One new promising area of science communication research in overcoming identity protective cognition is focusing on the salience of different identities that people hold. While much of the research on polarization of science has examined group identity, such as partisanship, priming relational identities, such as those of a parent or a friend, may pull people out of identity protective states. A recent study[71] tested the hypothesis that strategic communication can be used to prime different identities in people and that these different identities subsequently influence interpretations of messaging about politicized issues. This ultimately impacts beliefs, support for policies, and behaviors.[71] Focusing on the COVID-19 pandemic, Zeng[71] primed one group of participants in her study to have their relational identity of being a parent made salient and compared them to a control group who did not have any identity primed. When no identity was primed, people processed information about the COVID-19 pandemic through a partisan lens. However, when their parental identity was primed such partisan-based identity protective cognition of COVID-19 risk perceptions, protective behaviors, and policy support was reduced.

While the Zeng[71] study is novel and more research needs to be conducted on the influence of priming different identities to overcome identity protective cognition, this study is informative in that a focus on relational identity (parent, friend, etc.) may prove to be a useful tool to foster public engagement. Here, we return to our earlier assertions for knowledge engagement and community-based participatory research. In highlighting relevant relational identities for effective public engagement, scientists may come to value and respect communication *for* and *by* members of the public they seek to engage. Activating relatively subtle relational identities to overcome rigid identity protective states will require collaboration and partnership with all publics, but most especially those that are marginalized and underrepresented in scientific communities.

6.2 Abstract versus concrete cognition

When communicating about CRISPR, those facilitating public engagement should avoid abstract language and use concrete language to foster concrete cognition. "CRISPR" and "gene editing" are abstract terms to most people which can activate different understandings of what CRISPR is based on a person's mental schema. Using abstract language leads people to fill in information based on assumptions driven by their cognitive schema, which may be infected with misinformation and misperceptions about what CRISPR is. On the other hand, concrete language directs cognition to focus on specifics of the messages and has an advantage over abstract language for fast and accurate information processing. "Compared to abstract language, concrete language can be used to guide individuals out of their default cognitive schema to process messages in a new context, which then can in turn reshape their schema."[72] People's cognitive schemas are based on their own experiences.

Consider the following example of three science communicators–one is facilitating public engagement on CRISPR-based advancements in treating HIV, another is working on outreach for public support for implementing a CRISPR-based gene drive in mosquitos, and the third science communicator is concerned with facilitating public dialogue on the ethical considerations around CRISPR-based human germline editing. If these science communicators were asked to provide their thoughts and opinions on CRISPR in an abstract sense, the three of them would default to their own schema and their experiences that they have had communicating CRISPR. But all three have different schemas given the focus of their work and their different experiences. Now consider a layperson who has a very limited understanding of CRISPR. When asked about their opinions on CRISPR in an abstract sense, they will also rely on their mental schema, albeit quite fuzzy. Their mental schema might have been shaped by the news coverage of the He Jiankui scandal, which influences how they respond to abstract language about CRISPR. If CRISPR becomes heavily politicized, people's mental schema will begin to tie the biotechnology to their political identity which will foster identity protective cognition, motivated reasoning, and confirmation bias.

In a recent study, Tallapragada and her colleagues[72] examined the impact of abstract versus concrete information on attitudes toward genetically modified organisms (GMO). In a randomized controlled experiment, participants in the Tallapragada et al.[72] study were assigned to conditions that differed in abstract or concrete conceptualization of GMOs. Participants in the abstract conditions assessed risks and benefits of general "plant-based" and "animal-based" GMOs presented as such, whereas participants in the concrete conditions assessed risk and benefits of specific applications, such as "tomatoes with high levels of cancer fighting antioxidants" and "sheep whose milk can be used to produce medicines and vaccines."[72] Compared to participants in the abstract condition, participants in the concrete condition reported more positive affect toward GMOs, which in turn lead to greater support for specific GM applications and their intention to buy GM products. Using concrete language to pull participants away from defaulting on their mental schema and fostering the evaluation of specific GM applications led to more thoughtful and reasoned cognitive processing. Communication of science, especially one as complex as CRISPR technologies, requires concrete language that specifies the function and purpose of each application.

The psychological process of people defaulting to their mental schema in response to abstract messaging complicates public engagement of CRISPR-based technologies, as residual GMO framing influences information processing of public engagement messages on CRISPR. Scheufele et al.[13] call this the "'ghost of GMO past' or the rising concern that certain CRISPR applications might trigger public opposition of this kind that genetically modified organisms (GMOs) triggered when they first entered the marketplace." The 'Frankenfood' framing of GM crops took hold, and many have adopted the ideas that GM crops are not safe, not natural and pose a risk to the environment. But because the CRISPR-based technologies have yet to become politicized and framed in misleading ways, "science communicators have an opportunity to decide how to frame CRISPR-edited crops. We must assume that this window of constructive communication may be narrow, before genome editing becomes politicized, regulations are solidified, and companies stake their claims."[73]

6.3 The LIVA model for science communication

Developed by Jamieson and Hardy,[74] the LIVA model for science communication can be used to help pull people out of identity protective cognition when confronted with scientific information and evidence. When processing information, people are goal driven. These goals can either be *directional* in motivation leading people to come to conclusions that are in line with their previously held beliefs or people may be motivated by *accuracy*.[69] Directional motivated reasoning is identity protective, likely to be emotional, and use heuristics whereas accuracy motivated reasoning triggers more engaged and unbiased processing of the information. As Bolsen and Druckman[55] summarize, "Instead of defending a prior belief, identity, or worldview, an accuracy motivation leads individuals to assess all available information objectively, even if it runs counter to one's existing beliefs or identities." For example, if messaging on a scientific issue is priming political identity, then that message will be evaluated through a politicized lens, be identity protective, and trigger directional motivated reasoning. On the other hand, the LIVA model of science communication can be used to push audiences away from direction motivation and activate accuracy motivations by using four common-sense communication practices:

Leveraging: Cite and leverage credible sources that audiences trust and are valued across the ideological spectrum. For example, in empirically testing the LIVA model when communicating scientific data of shrinking Arctic Sea ice, Hardy and Jamieson[74] highlighted that NASA satellites collected the data that showed a negative trend in Arctic sea ice extent. Since NASA enjoys high credibility among both Republicans and Democrats, it did not trigger partisan bias.

Involving: The LIVA model for science communication invites the audience into the data as opposed to just telling the audience of conclusions that can be made from the data. This can be done by simply asking a series of questions about the data as part of LIVA messaging structure. For example, in the one study testing the LIVA model in communicating data on anthropogenic global warming, respondents were asked to look at a chart of global land and surface temperatures from 1880 to 2015 as they differed from the 20th century average. The participants of the study were then asked to rate the accuracy of statements about the data displayed in the chart, e.g., "How accurate is it to say that 14 of the 15 warmest years on record have fallen in the first 15 years of this century?" Asking direct questions about the data invites audiences to engage with the data, triggers accuracy motivated reasoning, which then fosters objective cognitive processing of the data.

Visualizing: In order to invite the audience into the data, it is helpful to provide visualizations of the scientific evidence, data, or trend. Especially when used in conjunction with audience involvement of the data, clear and easy to digest visualizations can help foster accuracy motivations and unbiased cognitive processing of the scientific evidence. This allows audiences to draw their own objective inferences.

Analogies: Analogies are useful in making complex concepts understandable and relatable to audiences. Analogy "is a powerful cognitive mechanism that people use to make inferences and learn new abstractions."[75] The LIVA model of science communication uses analogies to "translate the less familiar concepts in the science into referents tied to the experiences of the audience, this comparison trope can play a powerful role in communication of science."[74]

7. Conclusion

This chapter focuses on three areas for public engagement and communicating CRISPR-based technologies: (1) Understanding what the public knows about CRISPR and what it means to different publics, (2) shared and community-based knowledge production across diverse publics, and (3) suggestions for communicating CRISPR in a politically polarized age. First, a review of public opinion research on CRISPR-based technologies reveals that we do not know enough about how different publics make sense of the technology, especially outside the United States. Despite the lack of quality data, we can infer from initial results that people are more supportive of therapeutic applications of CRISPR-based technologies than enhancement applications. In addition, there is some evidence suggesting that support for CRISPR-based applications differs by sociodemographic groups. Unfortunately, robust trends are hard to identify because of a lack of quantity and quality of global public opinion data. Most of the data about the public understanding of CRISPR come from the United States.[20] For many communities that will be affected by CRISPR-based technologies, there is simply no public opinion data. An integrated global effort for scientifically rigorous public opinion research on CRISPR-based technologies is desperately needed and a better understanding of global views on and understanding of CRISPR is imperative for effective public engagement.

Many of the CRISPR advancements will impact underserved or marginalized communities that have typically been excluded from the knowledge production process. This chapter recommends that community-based participatory research (CBPR) can be used as an inclusive framework of public engagement and knowledge production within these communities with regards to CRISPR. Such an approach emphasizes the importance of the inclusion of identity-based concerns in communities where public engagement on CRISPR will be facilitated. These concerns can only be identified and considered with effective and inclusive engagement within those communities and we offer CBPR as an effective approach to do so.

Because of the trends of the politicization of science, we suspect that CRISPR applications will also become politicized in the near future and we therefore offer three communicative strategies to overcome identity protective cognition, confirmation bias, and motivated reasoning: (1) Priming a relational identity like that of a parent, child, sibling, or friend can be used to motivate people from away interpreting issues through a politicized and partisan len and promote more objective cognitive processing of information. (2) Using concrete, instead of abstract language, about the CRISPR-based applications can also motivate objective and thoughtful reasoning, especially since using abstract language runs the risk of triggering other historically politicized science related issues. (3) The LIVA model for science communication offers four ways those wishing to engage in public engagement of CRISPR can pull people away from identity protective cognition and process science information with accuracy motivations instead of directional ones.

Overall, there is much work that needs to be done for effective public engagement of CRISPR. There needs to be more quality data on global understanding of CRISPR technologies and the data needs to be collected expediently because of how fast CRISPR is advancing. Diverse publics, particularly underrepresented, marginalized, and underserved communities, need to be included in the shared knowledge production. Implementations of CRISPR applications, ethical considerations, regulations, and policy decisions need to be informed by effective communication and engagement with diverse publics but, unfortunately, much foundational work is still needed before we can begin to effectively do so.

References

1. Temple University. *Groundbreaking Research at Temple Paves Way for First Trial of CRISPR-Based HIV Therapy in Human Patients*; 2021. Accessed October 20, 2021 https://medicine.temple.edu/news/groundbreaking-research-temple-paves-way-first-trial-crispr-based-hiv-therapy-human-patients.
2. Mancuso P, Chen C, Kaminski R, et al. CRISPR based editing of SIV proviral DNA in ART treated non-human primates. *Nat Commun*. 2020;11(1):6065. https://doi.org/10.1038/s41467-020-19821-7.

3. McCullough M. *Temple Researchers Take Another Step Toward Curing HIV*. The Philadelphia Inquirer; 2021. Accessed October 21, 2021. https://www.inquirer.com/health/temple-researchers-get-approval-use-gene-editing-for-hiv-aids-20210917.html?utm_source=email&utm_campaign=edit_social_share_email_traffic&utm_medium=email&utm_content=&utm_term=&int_promo.
4. Nuñez JK, Chen J, Pommier GC, et al. Genome-wide programmable transcriptional memory by CRISPR-based epigenome editing. *Cell*. 2021;184(9):2503–2519. e17 https://doi.org/10.1016/j.cell.2021.03.025.
5. Alvarez J. *New CRISPR Technology Offers Unrivaled Control of Epigenetic Inheritance*. ScienceDaily; 2021. Accessed October 21, 2021. https://www.sciencedaily.com/releases/2021/04/210416131923.htm.
6. Jamieson KH, Albarracín D. The relation between media consumption and misinformation at the outset of the SARS-CoV-2 pandemic in the US. *Harv Kennedy Sch Misinformation Rev*. 2020. https://doi.org/10.37016/mr-2020-012.
7. Enders AM, Uscinski JE, Klofstad C, Stoler J. The different forms of COVID-19 misinformation and their consequences. *Harv Kennedy Sch Misinformation Rev*. 2020. https://doi.org/10.37016/mr-2020-48.
8. Romer D, Jamieson KH. Conspiracy theories as barriers to controlling the spread of COVID-19 in the U.S. *Soc Sci Med*. 2020;263:113356. https://doi.org/10.1016/j.socscimed.2020.113356.
9. Merkley E, Loewen PJ. Anti-intellectualism and the mass public's response to the COVID-19 pandemic. *Nat Hum Behav*. 2021;5(6):706–715. https://doi.org/10.1038/s41562-021-01112-w.
10. Van Kessel P, Quinn D. *Both Republicans and Democrats Cite Masks as a Negative Effect of COVID-19, but for Very Different Reasons*. Pew Research Center; 2020. Accessed October 21, 2021. https://www.pewresearch.org/fact-tank/2020/10/29/both-republicans-and-democrats-cite-masks-as-a-negative-effect-of-covid-19-but-for-very-different-reasons/.
11. Reuters Fact Check. *Fact Check-the COVID-19 Vaccine Does Not Change Recipients' Genetic Makeup*. Reuters; 2021. Accessed October 21, 2021. https://www.reuters.com/article/factcheck-vaccine-dna/fact-check-the-covid-19-vaccine-does-not-change-recipients-genetic-makeup-idUSL2N2L23EM.
12. Brossard D, Belluck P, Gould F, Wirz CD. Promises and perils of gene drives: navigating the communication of complex, post-normal science. *Proc Natl Acad Sci USA*. 2019;116(16):7692–7697. https://doi.org/10.1073/pnas.1805874115.
13. Scheufele DA, Krause NM, Freiling I, Brossard D. What we know about effective public engagement on CRISPR and beyond. *Proc Natl Acad Sci USA*. 2021;118(22):e2004835117. https://doi.org/10.1073/pnas.2004835117.
14. Funtowicz SO, Ravetz JR. Science for the post-normal age. *Futures*. 1993;25(7):739–755. https://doi.org/10.1016/0016-3287(93)90022-L.
15. Gallo ME, Sarata AK, Sargent Jr JF, Cowan T. *Advanced Gene Editing: CRISPR-Cas9*. Congressional Research Service; 2018. Accessed October 21, 2021. https://crsreports.congress.gov/product/pdf/R/R44824/6.
16. Brinegar K, Yetisen A, Choi S, et al. The commercialization of genome-editing technologies. *Crit Rev Biotechnol*. 2017;37(7):924–932. https://doi.org/10.1080/07388551.2016.1271768.
17. Verified Market Research. *Genome Editing Market Size and Forecast*; 2021. Accessed October 21, 2021. https://www.verifiedmarketresearch.com/product/global-genome-editing-market-size-and-forecast-to-2025/.
18. Jennings K. *The Gene Editing Startup Raised $315 Million for a Next Generation Crispr Tool to Cure Rare Diseases*. Forbes; 2021. Accessed October 21, 2021. https://www.forbes.com/sites/katiejennings/2021/07/13/this-gene-editing-startup-raised-315-million-for-a-next-generation-crispr-tool-to-cure-rare-diseases/?sh=3ca01aa55630.

19. Khusf G. Chapter 16: An ethic for enhancing human performance through integrative technologies. In: Bainbridge WS, Roco MC, eds. *Managing Nano-Bio-Info-Cogno Innovations: Converging Technologies in Society*. Springer; 2006:225–278.
20. Howell EL, Yang S, Beets B, Brossard D, Scheufele DA, Xenos MA. What do we (not) know about global views of human gene editing? Insights and blind spots in the CRISPR era. *CRISPR J*. 2020;3(3):148–155. https://doi.org/10.1089/crispr.2020.0004.
21. Weisberg SM, Badgio D, Chatterjee A. A CRISPR new world: attitudes in the public toward innovations in human genetic modification. *Front Public Health*. 2017;5:117. https://doi.org/10.3389/fpubh.2017.00117.
22. McCaughey T, Sanfilippo PG, Gooden GEC, et al. A global social media survey of attitudes to human genome editing. *Cell Stem Cell*. 2016;18(5):569–572. https://doi.org/10.1016/j.stem.2016.04.011.
23. The National Academies of Sciences Engineering Medicine. *On Human Gene Editing: International Summit Statement*; 2015. Accessed October 21, 2021. https://www.nationalacademies.org/news/2015/12/on-human-gene-editing-international-summit-statement.
24. Scheufele DA, Xenos MA, Howell EL, Rose KM, Brossard D, Hardy BW. U.S. attitudes on human genome editing. *Science*. 2017;357(6351):553–554. https://doi.org/10.1126/science.aan3708.
25. Funk C, Heffron M. *Public Views of Gene Editing for Babies Depend on How It Would Be Used*. Pew Research Center; 2018. Accessed October 2021. https://www.pewresearch.org/science/2018/07/26/public-views-of-gene-editing-for-babies-depend-on-how-it-would-be-used/.
26. Gaskell G, Bard I, Allansdottir A, et al. Public views on gene editing and its uses. *Nat Biotechnol*. 2017;35(11):1021–1023. https://doi.org/10.1038/nbt.3958.
27. Wang J-H, Wang R, Lee JH, et al. Public attitudes toward gene therapy in China. *Mol Ther Meth Clin Dev*. 2017;6:40–42. https://doi.org/10.1016/j.omtm.2017.05.008.
28. Critchley C, Nicol D, Bruce G, Walshe J, Treleaven T, Tuch B. Predicting public attitudes toward gene editing of germlines: the impact of moral and hereditary concern in human and animal applications. *Front Genet*. 2019;9:704. https://doi.org/10.3389/fgene.2018.00704.
29. Delhove J, Osenk I, Prichard I, Donnelley M. Public acceptability of gene therapy and gene editing for human use: a systematic review. *Hum Gene Ther*. 2020;31(1–2):20–46. https://doi.org/10.1089/hum.2019.197.
30. McPhetres J, Jong J, Zuckerman M. Religious Americans have less positive attitude toward science, but this does not extend to other cultures. *Soc Psychol Personal Sci*. 2021;12(4):528–536. https://doi.org/10.1177/1948550620923239.
31. McPhetres J, Zuckerman M. Religiosity predicts negative attitudes towards science and lower levels of science literacy. Jong J, *PLoS One*. 2018;13(11): e0207125. https://doi.org/10.1371/journal.pone.0207125.
32. Chan E. Are the religious suspicious of science? Investigating religiosity, religious context, and orientations towards science. *Public Underst Sci*. 2018;27(8):967–984. https://doi.org/10.1177/0963662518781231.
33. Haga SB, Rosanbalm KD, Boles L, Tindall GM, Livingston TM, O'Daniel JM. Promoting public awareness in engagement in genome sciences. *J Genet Couns*. 2013;22(4):508–516. https://doi.org/10.1007/s10897-013-9577-3.
34. Kampourakis K, Reydon TAC, Patrinos GP, Strasser BJ. Genetics and society—Educating scientifically literate citizens: introduction to the thematic issue. *Sci Educ*. 2014;23(2):251–258. https://doi.org/10.1007/s11191-013-9659-5.
35. McCaughey T, Budden DM, Sanfilippo PG, et al. A need for better understanding is the major determinant for public perceptions of human gene editing. *Hum Gene Ther*. 2019;30(1):36–43. https://doi.org/10.1089/hum.2018.033.

36. Marcon A, Master Z, Ravitsky V, Caulfield T. CRISPR in the North American popular press. *Genet Med.* 2019;21(10):2184–2189. https://doi.org/10.1038/s41436-019-0482-5.
37. Iyengar S, Peters MD, Kinder DR. Experimental demonstrations of the "not-so-minimal" consequences of television news programs. *Am Polit Sci Rev.* 1982;76(4): 848–858. https://doi.org/10.1017/S000305540018966X.
38. Iyengar S, Kinder DR. *News that Matters: Televisions and American Opinion.* University of Chicago Press; 1987.
39. Cyranoski D, Ledford H. International outcry over genome-edited baby claim. *Nature.* 2018;563(7733):607–608. https://doi.org/10.1038/d41586-018-07545-0.
40. Zhang X, Chen A, Zhang W. Before and after the Chinese gene-edited human babies: multiple discourses of gene editing on social media. *Public Underst Sci.* 2021; 30(5):570–587. https://doi.org/10.1177/0963662520987754.
41. Watanabe D, Saito Y, Tsuda M, Ohsawa R. Increased awareness and decreased acceptance of genome-editing technology: the impact of the Chinese twin babies. Garzon S, *PLoS One.* 2020;15(9): e0238128. https://doi.org/10.1371/journal.pone.0238128.
42. Calabrese C, Ding J, Millam B, Barnett GA. The uproar over gene-edited babies: a semantic network analysis of CRISPR on Twitter. *Environ Commun.* 2020; 14(7):954–970. https://doi.org/10.1080/17524032.2019.1699135.
43. Müller M, Schneider M, Salathé M, Vayena E. Assessing public opinion on CRISPR-Cas9: combining crowdsourcing and deep learning. *J Med Internet Res.* 2020;22(8): e17830. https://doi.org/10.2196/17830.
44. Hartley S, Thizy D, Ledingham K, et al. Knowledge engagement in gene drive research for malaria control. Bonizzoni M, *PLoS Negl Trop Dis.* 2019;13(4): e0007233. https://doi.org/10.1371/journal.pntd.0007233.
45. Yeh W-H, Shubina-Oleinik O, Levy JM, et al. In vivo base editing restores sensory transduction and transiently improves auditory function in a mouse model of recessive deafness. *Sci Transl Med.* 2020;12(546): eaay9101. https://doi.org/10.1126/scitranslmed.aay9101.
46. Michie M, Allyse M. Gene modification therapies: views of parents of people with Down syndrome. *Genet Med.* 2019;21(2):487–492. https://doi.org/10.1038/s41436-018-0077-6.
47. Feliú-Mójer MI. Gene editing communication must center marginalized communities. *Environ Commun.* 2020;14(7):877–880. https://doi.org/10.1080/17524032.2020.1812274.
48. World Health Organization. *Malaria*; 2021. https://www.who.int/news-room/fact-sheets/detail/malaria. Accessed September 28, 2021.
49. Mshinda H, Killeen GF, Mukabana WR, Mathenge EM, Mboera LEG, Knols BGJ. Development of genetically modified mosquitoes in Africa. *Lancet Infect Dis.* 2004;4(5):264–265. https://doi.org/10.1016/S1473-3099(04)01000-X.
50. Oetzel JG, Wallerstein N, Duran B, et al. Impact of participatory health research: a test of the community-based participatory research conceptual model. *Biomed Res Int.* 2018;2018:1–12. https://doi.org/10.1155/2018/7281405.
51. Wallerstein N, Duran B. Community-based participatory research contributions to intervention research: the intersection of science and practice to improve health equity. *Am J Public Health.* 2010;100(S1):S40–S46. https://doi.org/10.2105/AJPH.2009.184036.
52. Katz S. *Why Deaf People Oppose Using Gene Editing to "cure?" Deafness.* Discover Magazine; 2020. https://www.discovermagazine.com/health/why-deaf-people-oppose-using-gene-editing-to-cure-deafness. Accessed September 28, 2021.
53. Bauman H-DL, Murray JJ, eds. *Deaf Gain: Raising the Stakes for Human Diversity.* Minneapolis: University of Minnesota Press; 2014.

54. Hudson M, Mead ATP, Chagné D, et al. Indigenous perspectives and gene editing in Aotearoa New Zealand. *Front Bioeng Biotechnol.* 2019;7:70. https://doi.org/10.3389/fbioe.2019.00070.
55. Bolsen T, Druckman JN. Counteracting the politicization of science. *J Commun.* 2015;65(5):745–769. https://doi.org/10.1111/jcom.12171.
56. Scheufele DA. Science communication as political communication. *Proc Natl Acad Sci USA.* 2014;111(suppl 4):13585–13592. https://doi.org/10.1073/pnas.1317516111.
57. McCright AM, Dunlap RE. Anti-reflexivity: the American conservative movement's success in undermining climate science and policy. *Theory Cult Soc.* 2010;27 (2–3):100–133. https://doi.org/10.1177/0263276409356001.
58. McCright AM, Dunlap RE. The politicization of climate change and polarization in the American public's views of global warming, 2001–2010. *Sociol Q.* 2011;52(2):155–194. https://doi.org/10.1111/j.1533-8525.2011.01198.x.
59. Drummond C, Fischhoff B. Individuals with greater science literacy and education have more polarized beliefs on controversial science topics. *Proc Natl Acad Sci USA.* 2017;114 (36):9587–9592. https://doi.org/10.1073/pnas.1704882114.
60. Hardy BW, Tallapragada M, Besley JC, Yuan S. The effects of the "war on science" frame on scientists' credibility. *Sci Commun.* 2019;41(1):90–112. https://doi.org/10.1177/1075547018822081.
61. Nisbet EC, Cooper KE, Garrett RK. The partisan brain: how dissonant science messages lead conservatives and liberals to (dis)trust science. *Ann Am Acad Pol Soc Sci.* 2015;658 (1):36–66. https://doi.org/10.1177/0002716214555474.
62. Hardy BW, Tallapragada M. The moderating role of interest in politics and news consumption in the relationship between political ideology and beliefs about science and scientists in the United States. *Group Process.* 2021;24(5):783–796. https://doi.org/10.1177/13684302211013913.
63. Tollefson J. Trump vs Clinton: worlds apart on science. *Nature.* 2016;535(7613): 473–474. https://doi.org/10.1038/535473a.
64. Cook J, Lewandowsky S. Rational irrationality: modeling climate change belief polarization using Bayesian networks. *Top Cogn Sci.* 2016;8(1):160–179. https://doi.org/10.1111/tops.12186.
65. Rutjens BT, Sutton RM, van der Lee R. Not all skepticism is equal: exploring the ideological antecedents of science acceptance and rejection. *Pers Soc Psychol Bull.* 2018;44 (3):384–405. https://doi.org/10.1177/0146167217741314.
66. Frankovic K. *Fewer Than Half of Republicans Now Support Requiring Childhood Vaccinations for Infectious Diseases.* YouGovAmerica; 2021. Accessed October 21, 2021. https://today.yougov.com/topics/politics/articles-reports/2021/10/13/support-requiring-child-vaccinations.
67. Weber L, Barry-Jester AM. *Over Half of States Have Rolled Back Public Health Powers in Pandemic.* Kaiser Health News; 2021. Accessed October 21, 2021. https://khn.org/news/article/over-half-of-states-have-rolled-back-public-health-powers-in-pandemic/.
68. Kahan DM, Braman D, Gastil J, Slovic P, Mertz CK. Culture and identity-protective cognition: explaining the white-male effect in risk perception. *J Empir Leg Stud.* 2007;4(3):465–505. https://doi.org/10.1111/j.1740-1461.2007.00097.x.
69. Kunda Z. The case for motivated reasoning. *Psychol Bull.* 1990;108(3):480–498. https://doi.org/10.1037/0033-2909.108.3.480.
70. Akin H, Hardy BW, Jamieson KH. Countering identity-protective responses to climate change data. *Environ Commun.* 2020;14(8):1111–1126. https://doi.org/10.1080/17524032.2020.1776359.
71. Zeng C. A relational identity-based solution to group polarization: can priming parental identity reduce the partisan gap in attitudes toward the COVID-19 pandemic. *Sci Commun.* 2021;43(6):687–718. https://doi.org/10.1177/10755470211036676.

72. Tallapragada M, Hardy BW, Lybrand E, Hallman WK. Impact of abstract versus concrete conceptualization of genetic modification (GM) technology on public perceptions. *Risk Anal.* 2021;41(6):976–991. https://doi.org/10.1111/risa.13591.
73. Doxzen K, Henderson H. Is this safe? Addressing societal concerns about CRISPR-edited foods without reinforcing GMO framing. *Environ Commun.* 2020;14(7): 865–871. https://doi.org/10.1080/17524032.2020.1811451.
74. Jamieson KH, Hardy BW. Leveraging scientific credibility about Arctic Sea ice trends in a polarized political environment. *Proc Natl Acad Sci USA.* 2014;111(suppl 4): 13598–13605. https://doi.org/10.1073/pnas.1320868111.
75. Gentner D, Holyoak KJ. Reasoning and learning by analogy: introduction. *Am Psychol.* 1997;52(2):32–34. https://doi.org/10.1037/0003-066X.52.1.32.

CHAPTER NINE

Moral conviction: A challenge in the age of science politicization

Robin Bayes*
Department of Political Science, Northwestern University, Evanston, IL, United States
*Corresponding author: e-mail address: robin.bayes@u.northwestern.edu

Contents

1. What is moral conviction? 196
2. Why does politicization matter for moral convictions? 198
3. How prominent are moral convictions in public opinion on science issues? 202
4. How might strategic communication create moral conviction? 205
5. Moral conviction: A challenge to the foundations of science? 208
References 210

Abstract

Social science research can help science practitioners understand why the public responds to scientific findings differentially—sometimes believing, sometimes not. Four decades of research finds that people interpret science in ways that make it easier to dismiss scientific findings or consensuses that go against specific attitudes, or positions in social and policy debates, that they wish to maintain. This may be the case especially when a person's position is a moral conviction—that is, it is not only their preferred position, but what they feel is the morally correct position. This chapter explores why moral conviction matters for understanding public response to scientific information in the age of politicization, where moral conviction comes from, and the ways in which it poses a challenge to the foundations of science.

Scientific activity plays an important role in societal knowledge creation, learning, and decision-making.[1] However, particularly in democracies such as the United States, the ability of science to play this role depends on public buy-in.[2,3] Yet, public buy-in to scientific information is far from guaranteed. Some recent commentators suggest there is a "war on science."[4,5] Scholars debate whether public trust in science is declining, among whom, and why.[6,7] Most recently, the COVID-19 pandemic has painted a stark picture of the dangers of public disbelief in and mistrust of science.

On the other hand, the type of polarization around scientific findings seen during the pandemic does not always occur. On many topics, public

Progress in Molecular Biology and Translational Science, Volume 188
ISSN 1877-1173
https://doi.org/10.1016/bs.pmbts.2021.11.003

science beliefs are not notably polarized.[8,9] Social science research can help science practitioners understand why the public responds to scientific findings differentially—sometimes believing, sometimes not. Here, four decades of research finds that people interpret science in ways that make it easier to dismiss scientific findings and consensuses that are non-congenial to what they would like to believe. This includes findings that are antagonistic to people's self-interest and esteem,[10] their partisan and ideological attachments,[8] their cultural values,[11] and, especially, policy attitudes and opinions that they want to maintain.[12–14]

This chapter centers on a specific type of policy opinion that people can bend their science beliefs to protect: moral convictions. A moral conviction is when a person's position in a policy debate is not only their preferred position, but what they feel is the morally correct position.[15] When people see their position on an issue as a matter of moral right and wrong, they are less likely to change their opinion and less open to information that challenges it.[16,17] They are more likely to bend facts to fit their position[18] and to undermine oppositional evidence.[19] Scientific findings are often used to argue for and against positions in divisive social and policy debates, debates in which holding opinions with moral conviction is common. Therefore, the dynamics around moral convictions are relevant to understanding how the public interfaces with science.

In the next section, this chapter introduces the concept of moral conviction. The subsequent section explains why moral conviction matters for understanding public response to scientific information and discusses why the politicization of science creates an environment of ambiguity that is conducive to allowing people's moral convictions to bend their interpretations of science. Then, this chapter provides preliminary pilot data on the incidence of moral conviction across several issues related to biomedicine and discusses how people may come to hold moral convictions on such issues. Finally, this chapter offers concluding thoughts for science practitioners on challenges that moral conviction poses to the foundations of science.

1. What is moral conviction?

Agreement on an explicit definition of morality has eluded researchers and philosophers alike.[20] But in a general sense, morality pertains to fundamental, principled ideas about what is acceptable and unacceptable, good and bad. It is well-known that morality can influence public opinion. For example, moral concerns are relevant for many individuals' positions on questions like: "Does the state have the right to take a life? Is gambling inherently evil? Should homosexuality be on a moral and legal par with heterosexuality?"[21 (p675)] In short, moral concerns drive many people's opinions in debates about divisive social issues.

The moral convictions approach[15,22] investigates whether opinions driven by moral concerns are qualitatively different than opinions that are not. This approach measures, by self-report, the degree to which people hold their attitudes on social and policy issues with moral conviction. When people hold an attitude with moral conviction, they perceive their position to be a matter of moral right and wrong. This is sometimes also called moral construal[23] or a moralized attitude.[24] In contrast, people with attitudes that are not morally convicted might see them as matters of social convention or personal preference. Any attitude may be held with or without moral conviction, with patterned variations possible between cultures, time periods, and individuals.[25] Most importantly for the study of public opinion, two individuals may hold the same opinion but think about it differently, as either a moral position or not. For example, one person may perceive their position on taking, or not taking, the COVID-19 vaccine as a matter of personal taste, while another may consider their position a moral matter.

In extant work, moral conviction is typically self-reported, using a face-valid measure that has faced a number of tests for construct validity and reliability.[26] On a Likert scale, people report the degree to which a given attitude is "a reflection of your core moral beliefs and convictions." This straightforward approach works because, despite the lack of scholarly consensus on a precise definition of morality, people "simply know what morality is, often without being able to concretely define exactly what it means."[20 (p299)] Whether or not a person is morally convicted about something is a metacognition that they can readily access and report.[15]

The reason why the distinction between moral convictions and other attitudes is important is because moral convictions generally exhibit distinct characteristics, regardless of the specific position or moral reasoning behind it. When individuals perceive their position to be a matter of fundamental moral right and wrong, they are less likely to change their position and less open to arguments that challenge it.[16,17] They are more resistant to compromise and more likely to see compromise as impossible.[27,28] They also believe their position is more objective[15]; when their position is held with moral conviction, people are less likely to describe it as an opinion and more likely to state that someone who disagreed with their position was "surely mistaken."[29,30]

Furthermore, when people hold moral convictions, they are more likely to bend facts to fit their position and to undermine oppositional evidence.[18,19] This is the characteristic that is most important for understanding public response to science in the face of moral conviction. Scientific information is often used as factual support when arguing for and against positions in divisive social and policy debates. Importantly, though, policy decision-making is never simply about facts. To inform policy decisions, scientific findings must necessarily be combined with value-laden judgments about what outcomes are socially desirable.[2] Moral convictions develop when the public perceives these judgments to involve, not only what is desirable, but also what is morally good and bad, acceptable and unacceptable.

Disagreements over what is desirable, and what is morally good, can sometimes spill over into disagreements about facts. As Dietz[1 (p14086)] observes, people have a propensity to be "suspicious of facts that are difficult to reconcile with [their] values." In other words, moral convictions may drive people to dismiss scientific findings that are used to argue for a policy position that they feel is morally wrong or unacceptable. The next section will discuss this process in more detail, and why the age of science politicization creates conducive conditions for it to occur.

2. Why does politicization matter for moral convictions?

The "politicization of science" has been a subject of much recent research on public reception of science.[7,31–37] Despite much scholarly interest in the topic, however, the term itself has eluded a clearly agreed upon definition to date, as debates about the ways in which science is an object or site of politics have created various conceptualizations that bear only a loose family resemblance to each other.[38] Regardless, much existing work is united by an interest in how science is misused and manipulated by politically and economically self-interested actors, and the resulting consequences for public beliefs, interpretations, and attitudes about science.

Given these shared concerns, it may be helpful to think of politicization as when political interests shape the presentation of science to the general public. Actors may misuse and misrepresent scientific findings to further political agendas and bolster a desired outcome, such as a policy goal or avoidance of regulation.[39,40] This definition has the benefit of being both broad and comprehensive, encompassing a popular narrower definition of "when actors emphasize the uncertainty inherent in the scientific process to cast doubt on a scientific consensus,"[35,37,40] but also other forms of politicization, including selectively citing scientific evidence,[36] discrediting some science and/or scientists as biased, incompetent, or untrustworthy,[41] and overemphasizing dissenting voices.[42] Science is vulnerable to such manipulation because it involves inherent uncertainty and competing understandings, even when the process is working correctly. This aspect of the scientific enterprise can be difficult to communicate to lay audiences.[31,39]

An important consequence of the politicization of science is creating doubt within the public about whether scientific information and science-based arguments can be trusted.[37 (p150)] In an environment of politicization, science-based arguments may lose credibility when invoked in social and policy debates, as people are suspicious of whether the science is being harnessed for political ends.[33] In fact, simply bringing politicization to the public's attention, by pointing out that the production and presentation of science is often shaped by political motives, increases anxiety about and reduces support for applications of science in society. Bolsen, Druckman, and Cook[34] find that exposure to a message about politicization causes individuals to report heightened anxiety about nuclear energy and oppose its expansion, while Bolsen and Druckman[35] replicate these findings for two other energy technologies, hydraulic fracturing and carbon nanotubes. While there are ways to counteract these effects, including pre-emptive warnings and visual imagery, these solutions face challenges for large-scale implementation.[35–37]

By exploiting, and sometimes misrepresenting, the uncertainty inherent in the scientific enterprise, politicization creates ambiguity surrounding the veracity and integrity of science. This, in turn, creates a hospitable environment for segments of the public to interpret scientific findings or consensuses in diverging ways—with some people believing it and others finding reason to dismiss it. Differing interpretations are not harmful per se. However, individuals have a tendency toward "biased assimilation" of scientific information, meaning they may accept or reject it based on what they would like to believe.[13] When this occurs on a widespread basis, public science beliefs will not converge on a scientific consensus or even a general agreement.

To be clear, this does not always occur. Kahan and colleagues[43 (p196)] state that, for most topics, people conform their beliefs to the best available scientific evidence in much the same way that they "seek and follow expert medical treatment when they are ill or expert auto-mechanic service when their car breaks down." However, in some cases, people come to perceive certain scientific findings to be incompatible with their commitments to personally important beliefs and attitudes. Usually, this is facilitated by strategic communication from social and political elites that point out this incompatibility and make it salient to large swaths of the public. In these cases, people will instead engage with scientific information in ways that protect their pre-existing beliefs and opinions. This includes choosing not to personally believe scientific facts even while recognizing that they are endorsed by a consensus of the scientific community.[8,44]

Evidence that individuals resist believing scientific information that goes against what they wish to believe predates the age of politicization. Beginning with a landmark study by Lord and colleagues,[12] four decades of research finds that people often engage with information so as to arrive at a personally desired conclusion. Early studies exposed experimental subjects to scientific evidence that was either coherent with, or in opposition to, a personally desired conclusion. When asked to evaluate the evidence, subjects who desired to disbelieve the evidence were indeed less likely to believe it.[10] For instance, relative to non-smokers, smokers were less convinced by scientific information about the negative health effects of smoking, all else equal. Similarly, women who were heavy caffeine consumers were more skeptical of an article about caffeine risks to women than their low-caffeine consumer counterparts, while men, for whom the article was not personally relevant, did not exhibit this pattern.

Beyond personally desired conclusions due to self-interest, more recent work has identified evidence of biased information assimilation when it comes to other important commitments. This includes people's partisan and ideological attachments,[8] their cultural values,[11] and specific policy attitudes and opinions that they want to maintain.[13,14] People seem less likely to believe scientific findings or a scientific consensus when it goes against conclusions that are desired for these reasons as well. This pattern has driven a growing literature, and substantial debate, over the theory of "motivated reasoning."[45–48] Though often used imprecisely, the term generally refers to directional motivations—i.e., when people are motivated to selectively seek, evaluate, and believe information so as to fulfill a certain goal, such as to protect a pre-existing belief or attitude.[49,50]

When those pre-existing attitudes are also moral convictions—that is, when the position is not only a person's preferred position, but what they feel is the morally correct position—people may put forth extra effort to protect them. Morally convicted attitudes are especially stable and resistant to change.[16,17] Therefore, to avoid undermining or changing a morally convicted attitude, people may instead reverse engineer their factual beliefs, seeking what is known as moral coherence.[51,52] When it comes to policy positions, this means holding factual beliefs about a policy's implications that cohere with whether they feel the policy is morally good or bad. As a concrete example, Liu and Ditto[18] asked whether individuals felt capital punishment was morally acceptable or unacceptable even if it deterred crime, and their factual belief about how effective it truly was at deterring crime. Those who found capital punishment unacceptable, even if it deterred crime, believed it was less effective at deterring crime than those who found it acceptable. Notably, this effect was greatest among those who held their attitudes with the strongest moral conviction. In other words, when someone holds an opinion with moral conviction, they are more likely to seek moral coherence—by rejecting scientific information that goes against their morally preferred position—than someone who holds the same opinion but without, or with less, moral conviction.

Moral coherence seeking is a phenomenon that can occur with or without the politicization of science. However, one consequence of politicization is creating ambiguity around whether science can be trusted, and this works to facilitate biased information assimilation. A large body of work in psychology demonstrates that individuals interpret ambiguous information so as to achieve coherence with a desired conclusion.[53,54] This occurs in domains ranging from self-assessment, where evaluating the self on traits with greater ambiguity allows for more self-serving bias,[55,56] to politics, where perceived ambiguity allows people to discount condemning evidence against, and decide ballot challenges in favor of, their own political party.[57,58] In short, unambiguous information provides a constraint on biased information assimilation, while ambiguity enables it.[59] As Kunda[10 (p490)] states, "[Individuals] will believe undesirable evidence if they cannot refute it, but they will refute it if they can." Therefore, the ambiguity around the veracity of scientific information that is created when science is politicized likely enflames the process whereby moral convictions guide people's responses to scientific information.

Furthermore, politicization and moral conviction are likely often to coincide. In other words, the same issues for which the science is most politicized are also likely to exhibit higher average moral conviction levels in the public. This is due to what communication researchers have called the science "issue attention cycle," which alternates between phases when decisions are made in the administrative policy arena by scientists and other technical experts, and phases when decisions are moved to the overtly political arena and garner attention from interest groups, the media, and the public.[60,61] While science issues generally remain in the administrative arena, actions by political elites can bring them into the political arena. Once an issue enters the political arena, public opinion and values become important drivers of decision-making, giving interested actors more incentive to engage in politicization of the science. Nisbet and colleagues[60] also find that, once in the political arena, the moral rhetoric around a science issue becomes more prominent.

Greater amounts of moral rhetoric are likely to result in greater levels of moral conviction in the public, for reasons discussed in the penultimate section of this chapter.

3. How prominent are moral convictions in public opinion on science issues?

To set the stakes for why science practitioners should care about moral convictions and their impact on public views of science, data are needed on how widespread moral convictions actually are within the public, across a number of social and policy debates that commonly invoke scientific findings. To date, however, there is a dearth of public opinion work that systematically investigates this question. Polling data are sparse and fragmented, limiting comparability across different samples. Existing academic evidence of moral construal is limited, and either qualitative[62,63] or exclusively focused on the issue of climate change.[64,65]

To begin to fill this gap, the author conducted a pilot study in September 2021 that is the first comprehensive effort to measure moral conviction across a number of science issues. The pilot study took the form of a public opinion survey of U.S. adults fielded on Amazon's Mechanical Turk, an online crowdsourcing platform in which registered workers over the age of 18 can perform online tasks for payment, including participation in survey research. Workers were screened to include only those residing in the United States. In total, 620 participants completed the survey, of which 572 (92%) correctly answered a question designed to test their attention to the survey. Only those who passed the attention check were ultimately included in the analyses.

The survey consisted of 10 blocks of questions, with each block asking about different social or policy debates in which science is commonly invoked to argue for particular positions. Issues were selected for inclusion in the survey in order to cover substantial variance on a number of dimensions that extant research suggests may be relevant to moral conviction, including salience, ability to evoke emotions, and polarization by religiosity and ideology. The sample was split in half, such that each participant was randomly shown 5 out of the 10 blocks.

In each block, participants reported their attitude on a policy question related to that issue on a 7-point Likert scale (1 = Completely disagree, 7 = Completely agree). The attitude items used are displayed in Table 1. All statements are modeled after those commonly examined in public

Table 1 Pilot survey attitude items.

Issue	**Attitude item**
Childhood vaccination	Parents should vaccinate their children for childhood diseases, such as measles, mumps, rubella and polio
Climate change	People should take action to reduce carbon emissions and combat climate change
Combating obesity	The U.S. should implement measures against unhealthy foods, such as a soda tax, to combat obesity
COVID-19 vaccine	People should get vaccinated to protect themselves and others from COVID-19
Gene editing	The government should increase investment into gene editing research as a way to repair genetic defects and prevent genetic diseases
Genetically modified food	Food producers should label all foods that contain genetically modified organisms
Marijuana legalization	The U.S. should legalize marijuana for recreational use in all states
Nuclear power	The U.S. should build more nuclear power plants to increase clean electricity generation
Same-sex adoption	Same-sex couples should be allowed to adopt children through the same process as heterosexual couples
Transgender women in sports	Transgender women, i.e., individuals assigned male at birth who identify as women, should be allowed to compete in women's sports

opinion polling on divisive issues in which scientific findings are cited to support one or both sides.

Upon reporting their attitude, participants then answered the main outcome of interest, the extent to which their reported attitude was a reflection of their core moral convictions, on a 5-point Likert scale (1 = Not at all, 5 = Very Much). Fig. 1 presents the distribution of answers across the moral conviction spectrum for each issue attitude, with issues most relevant for biomedicine highlighted in the first panel and other issues examined in the survey shown in the second panel.

This pilot study is only a preliminary look at the prevalence of moral conviction in public attitudes on science-related issues. While results from studies conducted on Amazon's Mechanical Turk can be informative and

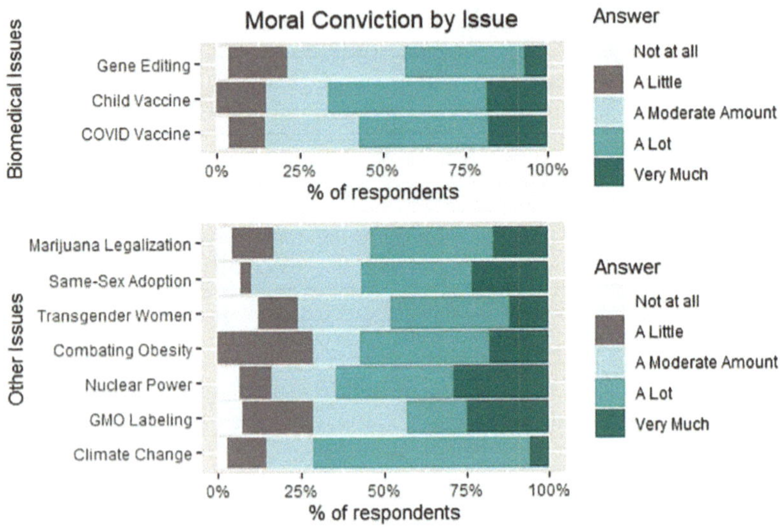

Fig. 1 Distribution of levels of moral conviction by issue attitude.

generalize fairly well to the American public at large, the platform does not provide a nationally representative sample.[66] To make precise inferences about moral conviction in the American public, the author plans to replicate this pilot study using a nationally representative sample that is large enough to investigate potential sub-group differences. This would provide additional helpful insights into who in the American public tends to hold morally convicted attitudes and on what issues.

However, even from this preliminary data, it is clear that many people do hold morally convicted attitudes on science issues. This is true even beyond issues that draw substantial attention in science communication literature, such as climate change, and issues that are extremely salient at the time of writing, such as the COVID-19 vaccine. Regarding the three issues most relevant for biomedicine—expanding investment into gene editing, vaccinating children for childhood diseases, and taking the COVID-19 vaccine—over three-quarters of respondents reported that their position on the issue reflected their core moral convictions at least "a moderate amount," if not more. In seven of the 10 issues included in the survey, over half of respondents reported that their position reflected their core moral convictions a lot or very much. The next section will discuss one way in which these moral convictions may develop within the public: through communication and framing.

4. How might strategic communication create moral conviction?

As stated above, moral rhetoric increases when science issues leave the more technical administrative arena and enter the more publicly visible political arena. In fact, as science becomes politicized, interested actors will often push issues into the political arena with the express purpose of strategically invoking morality: "The exclusion of certain moral values within the debates taking place in administrative settings...often forces opponents of scientific research to attempt to move debate to more overtly political arenas."[60 (p37)] Consequently, the strategic communication of moral rhetoric in political arenas likely leads people to think about science issues in a moral light for the first time. This strategic use of moral rhetoric is also known as moral framing or re-framing.[67] Moral framing refers to communication designed to increase the salience of moral considerations in the audience's understanding of an issue.

In general, framing is "the process by which people develop a particular conceptualization of an issue or reorient their thinking" about that issue.[68 (p104)] Communicators, both directly and through the media, will communicate using frames that help the public make sense of issues by providing interpretations of them.[69] The scientific and political discourse surrounding many social and policy issues is very complex, and most non-experts are not able or willing to invest a great deal of resources into acquiring the necessary information to understand it. Instead, they rely on guidance from trusted elite sources, including political parties, political leaders, and the media, to form their opinions.[70 (p446)] In giving this guidance, elite communicators will often strategically use framing to promote a certain conceptualization of the issue.

Framing might be especially important when it comes to communicating science to the public because concepts are new and largely unfamiliar[71,72] and because developing a fully informed opinion can require technical expertise that most of the public does not have.[73,74] Framing plays a crucial role in paring down the technical complexity of science issues, emphasizing some aspects of the issue over others and helping the public define the relevant considerations and sides.[75] Understanding framing in science communication encompasses understanding "how various actors in society define science-related issues in politically strategic ways, how journalists from various beats selectively cover these issues, and how diverse publics differentially perceive, understand, and participate in these debates."[76 (p1770)]

Although researchers have created an extensive typology of the various ways in which science-related issues are framed, the specific frame that is relevant here is the moral frame. This frame encourages the public to see science-related issues as a matter of moral right and wrong and provides cues that allow the public to reinforce the connection between moral considerations and particular issue positions.[61,77] As an example, Nisbet[77 (p21–22)] describes how Time magazine and Al Gore's WE campaign used moral framing to present climate change as a shared moral challenge, comparing climate action to efforts during the civil rights movement and World War II. Moral framing is commonly used by elites to communicate about science issues to the public.[60,61,78]

Theoretically, exposure to moral framing around any given issue should increase levels of moral conviction by reorienting people's thinking toward moral considerations. Two ways in which framing reorients people's thinking are by introducing new considerations into their understanding of an issue and reinforcing considerations that are already there so that they come to mind more easily.[68] The same frame may do one or the other, depending on whether the audience has ever encountered the considerations that are presented. As a concrete example, consider the following argument from the Danish Council of Ethics[79 (p9)] about the moral implications of the gene editing technique known as CRISPR:

> *CRISPR has "the potential to genetically enhance future human beings. If society cannot make such modifications available to all, the rich might pay to have their children modified to possess many of the traits associated with achieving success in our society. The competitive edge that these children already have will be sharpened further, and the gap between the top and bottom of society will grow wider with the potential risk of increasing existing inequalities within society as well between countries."*

In this argument, CRISPR is framed as a threat to the moral values of fairness and equity. For people who have never considered CRISPR in this light before, the frame will encourage them to integrate this new moral consideration into their understanding of the policy questions surrounding the use of CRISPR. For people who have encountered a similar frame before, this frame will act as a reminder, allowing this moral consideration to come to mind more easily whenever they think about the issue at a later time.

In theory, people's attitudes on issues are a function of the various considerations in their minds that are available, accessible, and applicable.[68] When moral frames introduce a new moral consideration about an issue to individuals, that consideration now becomes available to them. When moral frames repeat a moral consideration that the individual has already considered before, it now becomes more accessible and comes to mind more easily. Such frames will likely increase the audience's moral conviction because they will reorient their thinking to emphasize moral considerations. This should be the case regardless of whether the moral frame supports or challenges their specific position, because, even if the individual disagrees with the direction of the frame's argument, it will still encourage the individual to think about the issue as a matter of moral right and wrong.

There is also nascent empirical evidence on this point. Using three experiments centered on emergent issues from the area of food politics—natural food, factory farming, and genetically modified organisms (GMOs)—Clifford[80] finds that participants exposed to moral frames drawn directly from advocacy organizations like the Humane Society and PETA reported significantly stronger moralization of their position on the issue than control participants. In other words, exposure to moral framing can boost the level of moral conviction that people feel about their positions, at least for relatively unfamiliar issues.

It is important to note that there is also a growing body of suggestive and experimental evidence that moral framing works to change people's opinions.[46,78,81,82] For instance, Clifford and Jerit[61] identify the volume of moral arguments about stem cell research in the New York Times, then match it with a Pew survey of public attitudes on stem cell research in the same time period. They find that, as the volume of pro-stem cell moral arguments appearing in the news in the past 7 days increases, opinions in the survey are more favorable; similarly, as the volume of anti-stem cell moral arguments increases, opinions are less favorable. The persuasiveness of moral framing has also been demonstrated experimentally over a wide range of topics such as the environment, climate change, same-sex marriage, universal health care, military spending, and adopting English as the official language of the U.S.[67,83] Messages designed to move attitudes in a given direction by highlighting supportive moral considerations, e.g., in a pro-environmental direction by highlighting the need to protect the purity of the Earth, have been shown to be effective in creating opinion change.[81]

In short, there is growing evidence that moral framing is persuasive.[67] This is important for science practitioners to be aware of because, as interested parties push science issues from the administrative arena into the political arena, and advocates on both sides of any issue lean more on moral framing to advance their policy agendas, this will introduce more moral considerations into public understanding of science issues. As a result, morally convicted attitudes may become more common among individuals. This, in turn, will exacerbate the challenges associated with communicating non-congenial scientific findings to the public in an age of politicization. Some of these challenges even may pose a threat to the foundations of science.

5. Moral conviction: A challenge to the foundations of science?

In today's fragmented information environment, greater choice of information sources has undermined public consensus about the trusted gatekeepers of information.[84] As a result, science practitioners and communicators, who have long held cultural authority as privileged sources of information in most societies, must grapple with a flood of competing sources, as well as competing messaging that extends beyond communicating "just the facts." This environment has created space for the growing politicization of science and the resulting ambiguity among the public about what science to believe.

Contestation over science beliefs is central to the politicization of science, as actors shape "the presentation of scientific facts to fit distinct models of 'reality' for [politically] self-interested reasons."[34 (p4)] As a result, many scholars are concerned about politicization for its ability to promote science misperceptions, i.e., incorrect beliefs about the degree of scientific consensus on an issue, as well as personal beliefs that diverge from that consensus.[32,36,37,85] Because science produces knowledge that is the basis of many important societal systems and decisions, widely held misperceptions, or beliefs that do not reflect the best available scientific knowledge, can be problematic. These types of science misperceptions can be quite resistant to correction[86,87] and are especially so when motivated reasoning is involved.[88] Widespread misperceptions can result in sub-optimal decisions and, as starkly exemplified by the politicization of mask-wearing and other scientific recommendations during the COVID-19 pandemic, it can exacerbate dangers during times of crisis. At its most extreme, politicization may destabilize shared bases of knowledge so much that the very existence of reality is challenged, in what some have dubbed a "post-truth" era.[89] While contestation over facts, and anxiety about the status of knowledge within the public, are nothing new, the politicization of science necessarily brings these questions to the fore.[90]

Against this backdrop, it is important to understand the forces that lead people to take advantage of the ambiguity provided by politicization to interpret scientific findings in different ways. This chapter has discussed one such force, moral conviction, and explored how individuals with moral conviction will bend their factual beliefs to seek moral coherence. The resulting biased information assimilation may promote the development of incorrect beliefs in particular segments of the public, for whom scientific findings or consensuses threaten personally important beliefs and attitudes. However, it is not only important that individuals with morally convicted attitudes seek moral coherence; it is also important *how* they might go about doing so. Here, moral convictions may pose an even greater challenge to the foundations of science than promoting incorrect beliefs.

When faced with a scientific study that is non-congenial to their moral convictions, individuals seek moral coherence by searching for reasons to dismiss those findings. Two ways that people have been shown to justify dismissing non-congenial scientific findings are to evaluate the scientists who produced the findings, i.e., the source of the non-congenial information, as less authoritative[91,92] and to evaluate the study itself as flawed or poorly conducted.[12,13,57,93] In both cases, the process of seeking moral coherence results in the erosion of the credibility of scientists and the scientific method. Nauroth and colleagues[91 (p755)] explain the stakes: "If psychological biases do not only lead to disagreement about the matter of fact but entailed distrust and doubts about the competence and reputation of scientists...only because these scientists advocate undesirable research findings, a constructive dialogue [between scientists and laypeople] may be significantly impaired." Even eroding credibility in a few science domains is concerning, given findings that suggest that rejection of scientific authority can also "spillover" into uncontested spheres.[94]

This prospect is deeply troubling because the legitimacy allocated to science as a knowledge authority is what enables us to reap the social returns to scientific research.[2] Although science is not the only approach to systematic knowledge-making, it is a deeply established and culturally privileged one. As a result, most "knowledge societies" around the world run on expert processes based in science, and science plays a key role in informing societal and policy decision-making by providing information about how the world works.[95] Such social learning, Dietz[1] argues, is crucial for all societies to successfully adapt to stressors and avoid collapse.

It is clear, then, that the stakes could not be higher for science practitioners to understand how to communicate about scientific activity and maintain public buy-in to science in an age of politicization. This chapter has reviewed the ways in which moral conviction is a key factor in how the public forms beliefs and attitudes surrounding science. Moral conviction is already moderately high across a variety of social and policy issues that invoke science, and may grow higher, if the use of moral framing in public debates increases in the wake of the growing consensus about its persuasiveness. While this may pose a challenge to laypeople's foundational beliefs about the trustworthiness of scientific knowledge, research on these dynamics is in its infancy. This is an important avenue of future research on the "science of science."

References

1. Dietz T. Bringing values and deliberation to science communication. *Proc Natl Acad Sci.* 2013;110(suppl 3):14081–14087. https://doi.org/10.1073/pnas.1212740110.
2. Druckman J. Communicating policy-relevant science. *PS Polit Sci Polit.* 2015;48 (S1):58–69. https://doi.org/10.1017/s1049096515000438.
3. Shapin S. Cordelia's love: credibility and the social studies of science. *Perspect Sci.* 1995;3 (3):255–275.
4. Hardy B, Tallapragada M, Besley J, Yuan S. The effects of the "war on science" frame on scientists' credibility. *Sci Commun.* 2019;41(1):90–112. https://doi.org/10.1177/1075547018822081.
5. Mooney C. *The Republican War on Science.* New York: Basic Books; 2007.
6. Cofnas N, Carl N. Woodley of Menie M. does activism in social science explain conservatives' distrust of scientists? *Am Sociol.* 2018;49(1):135–148. https://doi.org/10.1007/s12108-017-9362-0.
7. Gauchat G. Politicization of science in the public sphere. *Am Sociol Rev.* 2012;77 (2):167–187. https://doi.org/10.1177/0003122412438225.
8. Kahan D. Climate-science communication and the measurement problem. *Polit Psychol.* 2015;36:1–43. https://doi.org/10.1111/pops.12244.
9. McPhetres J, Bago B, Pennycook G. Science beliefs, political ideology, and cognitive sophistication. *PSYARXIV.* 2021. https://osf.io/ad9v7/.
10. Kunda Z. The case for motivated reasoning. *Psychol Bull.* 1990;108(3):480–498. https://doi.org/10.1037/0033-2909.108.3.480.
11. Kahan D, Jenkins-Smith H, Braman D. Cultural cognition of scientific consensus. *J Risk Res.* 2011;14(2):147–174. https://doi.org/10.1080/13669877.2010.511246.
12. Lord C, Ross L, Lepper M. Biased assimilation and attitude polarization: the effects of prior theories on subsequently considered evidence. *J Pers Soc Psychol.* 1979;37 (11):2098–2109. https://doi.org/10.1037/0022-3514.37.11.2098.
13. Munro G, Ditto P. Biased assimilation, attitude polarization, and affect in reactions tostereotype-relevant scientific information. *Pers Soc Psychol Bull.* 1997;23 (6):636–653. https://doi.org/10.1177/0146167297236007.
14. Washburn A, Skitka L. Science denial across the political divide: liberals and conservatives are similarly motivated to deny attitude-inconsistent science. *Soc Psychol Personal Sci.* 2017;9(8):972–980. https://doi.org/10.1177/1948550617731500.
15. Skitka L, Hanson B, Morgan G, Wisneski D. The psychology of moral conviction. *Annu Rev Psychol.* 2021;72(1):347–366. https://doi.org/10.1146/annurev-psych-063020-030612.

16. Luttrell A, Togans L. The stability of moralized attitudes over time. *Pers Soc Psychol Bull.* 2021;47(4):551–564. https://doi.org/10.1177/0146167220935737.
17. Stanley M, Dougherty A, Yang B, Henne P, De Brigard F. Reasons probably won't change your mind: the role of reasons in revising moral decisions. *J Exp Psychol Gen.* 2018;147(7):962–987. https://doi.org/10.1037/xge0000368.
18. Liu B, Ditto P. What dilemma? Moral evaluation shapes factual belief. *Soc Psychol Personal Sci.* 2013;4(3):316–323. https://doi.org/10.1177/1948550612456045.
19. Bender J, Rothmund T, Nauroth P, Gollwitzer M. How moral threat shapes Laypersons' engagement with science. *Pers Soc Psychol Bull.* 2016;42(12):1723–1735. https://doi.org/10.1177/0146167216671518.
20. Skitka LJ, Conway P. Morality. In: Finkel EJ, Baumeister RF, eds. *Advanced Social Psychology: The State of the Science.* Oxford University Press; 2019:299–323.
21. Mooney C. The politics of morality policy: symposium editor's introduction. *Policy Stud J.* 1999;27(4):675–680. https://doi.org/10.1111/j.1541-0072.1999.tb01995.x.
22. Skitka L, Morgan G. The social and political implications of moral conviction. *Polit Psychol.* 2014;35:95–110. https://doi.org/10.1111/pops.12166.
23. Van Bavel J, Packer D, Haas I, Cunningham W. The importance of moral construal: moral versus non-moral construal elicits faster, more extreme, universal evaluations of the same actions. *PLoS One.* 2012;7(11):e48693. https://doi.org/10.1371/journal.pone.0048693.
24. Rhee J, Schein C, Bastian B. The what, how, and why of moralization: a review of current definitions, methods, and evidence in moralization research. *Soc Personal Psychol Compass.* 2019;13(12). https://doi.org/10.1111/spc3.12511.
25. Rozin P. The process of moralization. *Psychol Sci.* 1999;10(3):218–221. https://doi.org/10.1111/1467-9280.00139.
26. Skitka L. The psychology of moral conviction. *Soc Personal Psychol Compass.* 2010;4(4):267–281. https://doi.org/10.1111/j.1751-9004.2010.00254.x.
27. Mooney C, Schuldt R. Does morality policy exist? Testing a basic assumption. *Policy Stud J.* 2008;36(2):199–218. https://doi.org/10.1111/j.1541-0072.2008.00262.x.
28. Ryan T. No compromise: political consequences of moralized attitudes. *Am J Pol Sci.* 2017;61(2):409–423. https://doi.org/10.1111/ajps.12248.
29. Goodwin G, Darley J. The psychology of meta-ethics: exploring objectivism. *Cognition.* 2008;106(3):1339–1366. https://doi.org/10.1016/j.cognition.2007.06.007.
30. Goodwin G, Darley J. Why are some moral beliefs perceived to be more objective than others? *J Exp Soc Psychol.* 2012;48(1):250–256. https://doi.org/10.1016/j.jesp.2011.08.006.
31. Sarewitz D. How science makes environmental controversies worse. *Environ Sci Policy.* 2004;7(5):385–403. https://doi.org/10.1016/j.envsci.2004.06.001.
32. McCright A, Dunlap R. The politicization of climate change and polarization in the American public's views of global warming, 2001–2010. *Sociol Q.* 2011;52(2):155–194. https://doi.org/10.1111/j.1533-8525.2011.01198.x.
33. Lupia A. Communicating science in politicized environments. *Proc Natl Acad Sci.* 2013;110(suppl 3):14048–14054. https://doi.org/10.1073/pnas.1212726110.
34. Bolsen T, Druckman J, Cook F. How frames can undermine support for scientific adaptations: politicization and the status-quo bias. *Public Opin Q.* 2014;78(1):1–26. https://doi.org/10.1093/poq/nft044.
35. Bolsen T, Druckman J. Counteracting the politicization of science. *J Commun.* 2015;65(5):745–769. https://doi.org/10.1111/jcom.12171.
36. Bolsen T, Druckman J. Do partisanship and politicization undermine the impact of a scientific consensus message about climate change? *Group Process Intergroup Relat.* 2018;21(3):389–402. https://doi.org/10.1177/1368430217737855.

37. Bolsen T, Palm R, Kingsland J. Counteracting climate science politicization with effective frames and imagery. *Sci Commun.* 2019;41(2):147–171. https://doi.org/10.1177/1075547019834565.
38. Brown M. Politicizing science: conceptions of politics in science and technology studies. *Soc Stud Sci.* 2014;45(1):3–30. https://doi.org/10.1177/0306312714556694.
39. Druckman J. The crisis of politicization within and beyond science. *Nat Hum Behav.* 2017;1(9):615–617. https://doi.org/10.1038/s41562-017-0183-5.
40. Oreskes N, Conway E. *Merchants of Doubt.* USA: Bloomsbury Publishing; 2011.
41. Cann H, Raymond L. Does climate denialism still matter? The prevalence of alternative frames in opposition to climate policy. *Env Polit.* 2018;27(3):433–454. https://doi.org/10.1080/09644016.2018.1439353.
42. Clarke C. A question of balance. *Sci Commun.* 2008;30(1):77–107. https://doi.org/10.1177/1075547008320262.
43. Kahan D, Jenkins-Smith H, Tarantola T, Silva C, Braman D. Geoengineering and climate change polarization. *Ann Am Acad Pol Soc Sci.* 2015;658(1):192–222. https://doi.org/10.1177/0002716214559002.
44. Pasek J. It's not my consensus: motivated reasoning and the sources of scientific illiteracy. *Public Underst Sci.* 2018;27(7):787–806. https://doi.org/10.1177/0963662517733681.
45. Bayes R, Druckman JN. Motivated reasoning and climate change. *Curr Opin Behav Sci.* 2021;42:27–35. https://doi.org/10.1016/j.cobeha.2021.02.009.
46. Bayes R, Druckman JN, Goods A, Molden DC. When and how different motives can drive motivated political reasoning. *Polit Psychol.* 2020;41(5):1031–1052. https://doi.org/10.1111/pops.12663.
47. Hart P, Nisbet E. Boomerang effects in science communication. *Communic Res.* 2012;39(6):701–723. https://doi.org/10.1177/0093650211416646.
48. Tappin B, Pennycook G, Rand D. Thinking clearly about causal inferences of politically motivated reasoning: why paradigmatic study designs often undermine causal inference. *Curr Opin Behav Sci.* 2020;34:81–87. https://doi.org/10.1016/j.cobeha.2020.01.003.
49. Druckman J, McGrath M. The evidence for motivated reasoning in climate change preference formation. *Nat Clim Chang.* 2019;9(2):111–119. https://doi.org/10.1038/s41558-018-0360-1.
50. Molden DC, Higgins ET. Motivated thinking. In: Holyoak KJ, Morrison RG, eds. *The Oxford Handbook of Thinking and Reasoning.* Oxford University Press; 2012:390–409.
51. Clark C, Chen E, Ditto P. Moral coherence processes: constructing culpability and consequences. *Curr Opin Psychol.* 2015;6:123–128. https://doi.org/10.1016/j.copsyc.2015.07.016.
52. Ditto PH, Liu BS. Moral coherence and political conflict. In: Valdesolo P, Graham J, eds. *Social Psychology of Political Polarization.* Routledge; 2016:102–122.
53. Holyoak K, Simon D. Bidirectional reasoning in decision making by constraint satisfaction. *J Exp Psychol Gen.* 1999;128(1):3–31. https://doi.org/10.1037/0096-3445.128.1.3.
54. Simon D, Stenstrom D, Read S. The coherence effect: blending cold and hot cognitions. *J Pers Soc Psychol.* 2015;109(3):369–394. https://doi.org/10.1037/pspa0000029.
55. Dunning D, Meyerowitz J, Holzberg A. Ambiguity and self-evaluation: the role of idiosyncratic trait definitions in self-serving assessments of ability. *J Pers Soc Psychol.* 1989;57(6):1082–1090. https://doi.org/10.1037/0022-3514.57.6.1082.
56. Logg J, Haran U, Moore D. Is overconfidence a motivated bias? Experimental evidence. *J Exp Psychol Gen.* 2018;147(10):1445–1465. https://doi.org/10.1037/xge0000500.
57. Munro G, Munro C. "Soft" versus "hard" psychological science: biased evaluations of scientific evidence that threatens or supports a strongly held political identity. *Basic Appl Soc Psych.* 2014;36(6):533–543. https://doi.org/10.1080/01973533.2014.960080.
58. Kopko K, Bryner S, Budziak J, Devine C, Nawara S. In the eye of the beholder? Motivated reasoning in disputed elections. *Polit Behav.* 2011;33(2):271–290. https://doi.org/10.1007/s11109-010-9133-x.

59. Leeper T, Slothuus R. Political parties, motivated reasoning, and public opinion formation. *Polit Psychol.* 2014;35:129–156. https://doi.org/10.1111/pops.12164.
60. Nisbet M, Brossard D, Kroepsch A. Framing science. *Harv Int J Press Polit.* 2003;8 (2):36–70. https://doi.org/10.1177/1081180x02251047.
61. Clifford S, Jerit J. How words do the work of politics: moral foundations theory and the debate over stem cell research. *J Theor Polit.* 2013;75(3):659–671. https://doi.org/10.1017/s0022381613000492.
62. Sadler T, Zeidler D. The morality of socioscientific issues: construal and resolution of genetic engineering dilemmas. *Sci Educ.* 2004;88(1):4–27. https://doi.org/10.1002/sce.10101.
63. Whitmarsh L. What's in a name? Commonalities and differences in public understanding of "climate change" and "global warming". *Public Underst Sci.* 2009;18(4):401–420. https://doi.org/10.1177/0963662506073088.
64. Doran R, Böhm G, Pfister H, Steentjes K, Pidgeon N. Consequence evaluations and moral concerns about climate change: insights from nationally representative surveys across four European countries. *J Risk Res.* 2019;22(5):610–626. https://doi.org/10.1080/13669877.2018.1473468.
65. Markowitz E. Is climate change an ethical issue? Examining young adults' beliefs about climate and morality. *Clim Change.* 2012;114(3–4):479–495. https://doi.org/10.1007/s10584-012-0422-8.
66. Mullinix K, Leeper T, Druckman J, Freese J. The generalizability of survey experiments. *J Exp Polit Sci.* 2015;2(2):109–138. https://doi.org/10.1017/xps.2015.19.
67. Feinberg M, Willer R. Moral reframing: a technique for effective and persuasive communication across political divides. *Soc Personal Psychol Compass.* 2019;13(12). https://doi.org/10.1111/spc3.12501.
68. Chong D, Druckman J. Framing theory. *Annu Rev Polit Sci.* 2007;10(1):103–126. https://doi.org/10.1146/annurev.polisci.10.072805.103054.
69. Gamson W, Modigliani A. Media discourse and public opinion on nuclear power: a constructionist approach. *Am J Sociol.* 1989;95(1):1–37. https://doi.org/10.1086/229213.
70. Rugeley C, Gerlach J. Understanding environmental public opinion by dimension: how heuristic processing mitigates high information costs on complex issues. *Polit Polic.* 2012;40(3):444–470. https://doi.org/10.1111/j.1747-1346.2012.00352.x.
71. Cobb M. Framing effects on public opinion about nanotechnology. *Sci Commun.* 2005;27(2):221–239. https://doi.org/10.1177/1075547005281473.
72. Druckman J, Bolsen T. Framing, motivated reasoning, and opinions about emergent technologies. *J Commun.* 2011;61(4):659–688. https://doi.org/10.1111/j.1460-2466.2011.01562.x.
73. Carmichael J, Brulle R. Elite cues, media coverage, and public concern: an integrated path analysis of public opinion on climate change, 2001–2013. *Env Polit.* 2017;26 (2):232–252. https://doi.org/10.1080/09644016.2016.1263433.
74. Weber E, Stern P. Public understanding of climate change in the United States. *Am Psychol.* 2011;66(4):315–328. https://doi.org/10.1037/a0023253.
75. Nisbet M, Scheufele D, Shanahan J, Moy P, Brossard D, Lewenstein B. Knowledge, reservations, or promise? *Commun Res.* 2002;29(5):584–608. https://doi.org/10.1177/009365002236196.
76. Nisbet M, Scheufele D. What's next for science communication? Promising directions and lingering distractions. *Am J Bot.* 2009;96(10):1767–1778. https://doi.org/10.3732/ajb.0900041.
77. Nisbet MC. Framing science: a new paradigm in public engagement. In: Kahlor L, Stout P, eds. *Understanding Science: New Agendas in Science Communication.* Routledge; 2009:40–67.

78. Clifford S, Jerit J, Rainey C, Motyl M. Moral concerns and policy attitudes: investigating the influence of elite rhetoric. *Polit Commun.* 2015;32(2):229–248. https://doi.org/10.1080/10584609.2014.944320.
79. The Danish Council on Ethics. *Statement From the Danish Council on Ethics on Genetic Modification of Future Humans*; 2021. Published 2016. Accessed November 10, 2021 https://www.etiskraad.dk/~/media/Etisk-Raad/en/Publications/Statement-on-genetic-modification-of-future-humans-2016.pdf.
80. Clifford S. How emotional frames moralize and polarize political attitudes. *Polit Psychol.* 2019;40(1):75–91. https://doi.org/10.1111/pops.12507.
81. Feinberg M, Willer R. The moral roots of environmental attitudes. *Psychol Sci.* 2013;24(1):56–62. https://doi.org/10.1177/0956797612449177.
82. Wolsko C, Ariceaga H, Seiden J. Red, white, and blue enough to be green: effects of moral framing on climate change attitudes and conservation behaviors. *J Exp Soc Psychol.* 2016;65:7–19. https://doi.org/10.1016/j.jesp.2016.02.005.
83. Feinberg M, Willer R. From gulf to bridge. *Pers Soc Psychol Bull.* 2015;41(12):1665–1681. https://doi.org/10.1177/0146167215607842.
84. Iyengar S, Massey D. Scientific communication in a post-truth society. *Proc Natl Acad Sci.* 2019;116(16):7656–7661. https://doi.org/10.1073/pnas.1805868115.
85. Jones-Jang S, Noland C. The politicization of health and science: role of political cues in shaping the beliefs of the vaccine-autism link. *Health Commun.* 2020;1–9. https://doi.org/10.1080/10410236.2020.1859723.
86. Nyhan B, Reifler J, Richey S, Freed G. Effective messages in vaccine promotion: a randomized trial. *Pediatrics.* 2014;133(4):e835–e842. https://doi.org/10.1542/peds.2013-2365.
87. Carey J, Chi V, Flynn D, Nyhan B, Zeitzoff T. The effects of corrective information about disease epidemics and outbreaks: evidence from Zika and yellow fever in Brazil. *Sci Adv.* 2020;6(5):1–10. https://doi.org/10.1126/sciadv.aaw7449.
88. van Stekelenburg A, Schaap G, Veling H, Buijzen M. Correcting misperceptions: the causal role of motivation in corrective science communication about vaccine and food safety. *Sci Commun.* 2020;42(1):31–60. https://doi.org/10.1177/1075547019898256.
89. McIntyre L. *Post-Truth.* MIT Press; 2018.
90. Jasanoff S, Simmet H. No funeral bells: public reason in a 'post-truth' age. *Soc Stud Sci.* 2017;47(5):751–770. https://doi.org/10.1177/0306312717731936.
91. Nauroth P, Gollwitzer M, Kozuchowski H, Bender J, Rothmund T. The effects of social identity threat and social identity affirmation on laypersons' perception of scientists. *Public Underst Sci.* 2017;26(7):754–770. https://doi.org/10.1177/0963662516631289.
92. Nisbet E, Cooper K, Garrett R. The partisan brain. *Ann Am Acad Pol Soc Sci.* 2015;658(1):36–66. https://doi.org/10.1177/0002716214555474.
93. Greitemeyer T. I am right, you are wrong: how biased assimilation increases the perceived gap between believers and skeptics of violent video game effects. *PLoS One.* 2014;9(4). https://doi.org/10.1371/journal.pone.0093440.
94. Roos J. Contested knowledge and spillover. *Soc Curr.* 2017;4(4):360–379. https://doi.org/10.1177/2329496516670187.
95. Knorr-Cetina K. *Epistemic Cultures.* Cambridge, MA: Harvard University Press; 2003.

CHAPTER TEN

Strategic science communication in the age of politicization

Todd P. Newman[a,*]
[a]Department of Life Sciences Communication, University of Wisconsin-Madison, Madison, WI, United States
*Corresponding author: e-mail address: tpnewman@wisc.edu

Contents

1. Introduction 215
2. Scientists attitudes and behavior toward public communication 217
   2.1 Goals for public engagement and communication 218
   2.2 Factors that predict willingness to engage 218
3. Strategic science communication: Connecting goals, objectives, and tactics 221
4. The brand of science and why it matters 223
5. Conclusion 227
Acknowledgments 228
References 228

Abstract

This chapter provides an overview of strategic communication concepts and frameworks that support scientists' communication and engagement efforts. While this chapter provides a synthesis of the research in these different areas of science communication and how the life sciences and medical science can leverage particularly insights, communicating science in an age of politicization requires a reflection not just at what drives differences in how the public thinks about the sciences and scientists, but also insights into our shared thoughts and feelings about science. To that end, the chapter concludes with a broader examination of the connection between science and society through an overview of new and innovative research on how both scientists and the public think and feel about science and the sciences more specifically. Connecting these different frameworks and concepts provides important directions for the life science community to consider in their specific communication and engagement efforts, and those of the scientific community more broadly.

1. Introduction

Over the last several decades, leaders from the scientific community have urged their colleagues to not only increase their frequency of communication with various publics, but also the scope of communication and

Progress in Molecular Biology and Translational Science, Volume 188
ISSN 1877-1173
https://doi.org/10.1016/bs.pmbts.2021.12.001

engagement activities.[1–3] This shift recognizes that the longstanding assumption that communication should be focused on increasing the public's understanding and excitement about science not only lacks empirical support as a means of changing attitudes or beliefs about science,[4] but also doesn't acknowledge the nature of "post-normal" science, in which societal implications of science are far reaching.[5] Now, communication efforts emphasize forms of communication based on engagement, where communication is bi-directional with different publics, and promotes public participation, mutual understanding, and relationship building based on shared values and beliefs.[6] The motivation for encouraging these behaviors stems from a desire among the scientific community to shape the public image of science, as well as ensure Americans support funding for scientific research, value the use of scientific evidence in decision-making, and overall, reinforce strong attitudes toward science and the scientific process.[3]

While communication and engagement has traditionally been viewed by scientists as outside the scope of their job, a growing number of scientists across academia, government, and the private sector see public communication and engagement as critical due to societal debates about science issues and the (mis)use of science in policy decision-making.[7,8] As scientific advances increasingly intersect with social, legal, and ethical dilemmas, they require scientists and the scientific community to understand the different goals that they have for public communication, and what specific strategies may be used to achieve these goals.[9,10] More will be discussed on this later in the chapter, but it is important to note the way in which science communication moved from a "one-size-fits-all" approach, to one that more carefully approaches communication through a strategic framework.[11]

The push for scientists increased communication and engagement rests on a strong foundation: Internationally, and specifically in the U.S., the public has a great deal of trust in scientists, overall, and views the benefits of science to outweigh the risks, overall. Compared to other institutions, levels of trust in science are one of the highest—trailing only behind the military—and have remained stable over time.[12] However, differences emerge when looking at trust in confidence of the scientific community and the priority of issues to address. Partisanship and level of education are just some of the factors that drive differences in attitudes toward what should be a priority for scientific research and their relative trust in different experts.[13] Trust in science and scientists, therefore, is much more nuanced than aggregate measures portray, often undermined by previous communications from the scientific community that fail to connect to specific audiences' value systems, and their view of proposed outcomes. Moreover, just as there is no "one-size-fits-all" approach to communication, the same can be said to science or scientists. Rather science constitutes various different *sciences* and *scientists*. In other words, while trust in science, in the aggregate, appears to strong among the US public, that trust is driven by the types of sciences and scientists that confirm what they want to hear.[14]

It is within this context that a growing field of research in science communication focused on surveying scientists to better understand how they approach public communication and engagement, and the factors that predict their willingness to engage. And while scientists have good intentions in their efforts to communication with public audiences, they also need to develop these skills and strategy. Science communication training is rapidly developing to help scientists develop these skills and strategies.[15] Today, science communication training courses are integrated into graduate STEM education, as well as through a growing number of research centers and organizations providing communication training to university and government-based scientists, scientists in private industry, as well as medical professionals.[16] This chapter provides an overview of the extent to which strategic communication concepts and frameworks integrate into scientists' communication and engagement efforts, and the role of communication training in supporting these efforts. Finally, while this chapter provides a synthesis of the research in these different areas of science communication and how the life sciences and medical science can leverage particularly insights, communicating science in an age of politicization requires a reflection not just at what drives differences in how the public thinks about the sciences and scientists, but also insights into our shared thoughts and feelings about science. To that end, the chapter concludes with a broader examination of the connection between science and society through an overview of new and innovative research on how both scientists and the public think and feel about science and the sciences more specifically. Connecting these different frameworks and concepts provides important directions for the life science community to consider in their specific communication and engagement efforts, and those of the scientific community more broadly.

2. Scientists attitudes and behavior toward public communication

As a result of the scientific communities continued calls for scientists to engage with various non-expert audiences, there is a growing body of research within the science communication literature examining the factors that influence scientists' willingness to engage.[11] Through a number of small- and large-scale surveys of scientists, scholars have utilized various frameworks to provide a theoretical lens to understand the factors that influence willingness to take part in public communication. The overall goal of this line of research is to provide a landscape of how scientists approach public communication, including their attitudes towards communication and their confidence in the ability to do so.

2.1 Goals for public engagement and communication

To begin, there are a number of potential goals for why a scientist might choose to engage in public communication activities. In a 2020 survey of scientists in the US and Canada, Besley et al.[10] measured scholars expressed goals for public communication. When asked about the broad goals for their communication, scientists indicate overall strong support for a number of diverse goals, including: "ensuring that policymakers use scientific evidence," "ensuring that our culture values science," "ensuring adequate funding for scientific research," "helping people use science to make better personal decisions," "fulfilling a duty to society," and "professional reputation." Across the sample, scientists ranked the goals of "ensuring our culture values science" and "ensuring policymakers use scientific evidence" the highest, with the goals of "professional reputation" and "helping people make more informed decisions" lower.

In a related, but more conceptual level, Scheufele et al.[9] outlined seven goals for public engagement more specifically in the context of emerging technology. These goals range in terms of how different publics might exert influence on policy or the scientific enterprise, including: "avoid potential controversy," "educate the public," "build democratic capacity through deliberation," "widen representation of voices," "solicit input on value debates," "enable responsible innovation," and "shape policy." In addition, the authors suggest several different types of modalities where these goals may be reached, including: communication, consultation, involvement, collaboration, and empowerment.

2.2 Factors that predict willingness to engage

While the motivation to engage may be influenced by a single goal or a combination of different goals, there are social, cultural, and demographic factors, that to a varying degree, either encourage of inhibit scientists' willingness to engage in public communication. First, there continues to be consistent evidence of a strong willingness among scientists for public engagement and communication activities across a range of contexts,[7,8,15] especially among scientists who indicate they had positive past experiences. This includes face-to-face engagement, engaging with policy-makers, being interviewed by journalists and other media professionals, online engagement, and to a lesser extent, protest or direct-advocacy. In terms of whether scientists from certain disciplines are more likely to engage than others, few differences among willingness to engage are found when looking across academic discipline. This partly suggests that the culture towards public communication and engagement is common across fields. This may come at somewhat of a surprise to many scientists who have taken part in public communication activities, as the advice often given is to connect your science to something that the audience values and what it means to them personally. Scientists, therefore, in more applied fields—such as the social sciences,[17] environmental science,[17] or biological/medical sciences,[17–19] are then perceived as more likely to engage than those who focus on more basic science research questions, such as in the fields of physics or chemistry, where the outcome of the research is not clear. While differences across academic disciplines may not drive any differences in terms of willingness to engage, differences in the types of research questions that scientists seek to address (basic vs applied), may unpack unique motivations about how they approach science communication and engagement. This will be further discussed in the chapter, as there are important differences that the scientific community needs to think carefully about.

While scientists overall indicate a strong willingness to engage in public communication, normative factors continue to have a mixed influence on scientists' willingness to engage. For example, while many scientists have a strong desire to engage in public communication, institutional norms, such as how their department, college, or university perceives public communication may influence a scientist's behavior. For example, many academic scientists indicate that public engagement and communication does not count towards tenure and promotion, thus inhibiting their motivation for communication and engagement. In their study of scientists at land grant universities in the US, Rose et al.[8] found that less than half (43%) of respondents thought high level administrators prioritized public communication. Likewise, only 23% thought that public communication was "highly important" to their colleagues, and only half thought that scientists who participate in public communication are well regarded by their peers.

Other studies find, although less consistent, that how they scientists perceive their colleagues' views on public engagement and communication strongly influences their own behavior. While scientists overall indicate mixed views on how public communication is perceived by their colleagues and institution, recent studies indicate that norms do not have a significant relationship with ones' willingness to engage.[7]

Finally, the most predictive factor influencing a scientist's willingness to engage is confidence in their communication skills. For the majority of scientists, communication and the social sciences more broadly, are not part of their training. While this trend is shifting as science communication graduate seminars, minors, and training or fellowships become more integrated into STEM education,[14] it demonstrates that communication, like any other skill, must be practiced. While there is no doubt that some scientists are naturally gifted communicators, for the majority, much like their willingness to engage, there is support for, and acknowledgement of, strengthening their communication skills. To that end, a number of (science) communication training programs are now active around the world, some as non-profit organizations, while others situated in institutes or departments across colleges and universities. A key goal for these organizations is to help scientists develop and sharpen their communication skills through audience-based approaches, such as storytelling and message design.[20] Through these efforts, there is growing, but still not sufficient understanding, that there is no "public," but rather various "publics." These publics vary based on their level of science literacy and knowledge, values, beliefs, and attitudes towards science, in general, and science-issues, specifically.

The effect of training on scientists' willingness to engage is only recently being examined empirically.[20–23] Previous studies found a positive relationship between training and public engagement participation,[24,25] while other research failed to find such relationship between training and participation in engagement activities.[26] Likewise, other studies found mixed results in testing whether scientists who received communication training are actually perceived as more effective communicators compared to those that did not receive training.[27] It is also possible that training may indirectly influence scientists' willingness to engage through psychological factors, i.e., scientists' self-perceived engagement abilities, and attitudes toward their audiences. Participating in science communication training has the potential to equip scientists with skills and confidence that are ultimately associated with greater willingness to engage with the public.[28] Moreover, these psychosocial factors are likely to be influenced by sociological factors. For instance, scientists' access to training opportunities can depend on university level-resources and normative beliefs about participating in public engagement with science.[29] The reason that communication training may have different effects is because they cover a wide range of goals and content, from building knowledge and skills to incentivize affective attachments to the public, to building one's identity as a science communicator.[30] More substantial research and evaluation, addressing a broader range of science communication training programs skills and agendas, is needed to better guide professionals in ways to support scientists in communicating with the public effectively.[31]

Understanding why a scientist chooses to engage in public communication is influence by a number of social, cultural, and individual-level factors, all of which, to an extent, either may encourage or inhibit a scientist from engaging. While institutional norms are found not to play a significant role in motivating a scientist to engage, for the majority, the greatest predictor of public communication is the scientist's confidence in their own communication skills and the belief that communication can have a beneficial effect.[7] To the extent that public communication may reach its intended goal, scientists need to take a more strategic orientation to science communication, where they recognize that different communication strategies—through the careful choice of communication objectives and tactics—help to reach a desired goal, but require different sets of skills.

3. Strategic science communication: Connecting goals, objectives, and tactics

Efforts to encourage scientists to think strategically about their public communication activities continues to grow within the science communication community. As scientists recognize that goals for communication extend beyond just increasing knowledge, but rather distinct goals that may lead to different outcomes, it is important to recognize that not all communication is a "one-size-fits-all" approach. In this context, strategic science communication is encouraging scientists to recognize how certain goals for public communication may be achieved through careful consideration of different communication objectives and communication tactics. Communication objectives are viewed as antecedent of goals, and help guide the eventual choice of tactics to use. Together, the goals-objectives-tactics framework provides communicators with a framework to think about their audience and what long-term goal that they want to achieve.[11]

Building off of the goals that were described previously, there are a number of communication objectives that the scientific community may prioritize in their communication efforts. The most common objectives in science communication historically are "ensuring people are informed about scientific issues" and "getting people interested and excited about science." While these have previously been conceptualized as goals, they are more suitable as tactics as they do not lead to an eventual outcome. Expanding on these objectives, Besley et al.[32] developed six additional objectives based on interviews with scientists and communication trainers. These include: "demonstrating expertise," "hearing what others think about scientific issues," "demonstrating care about societies well-being," "demonstrating openness and transparency," and "demonstrating shared community values." In a survey of academic scientists from five professional scientific societies, scientists indicate overall broad support for the different objectives. However, somewhat concerning is that "ensuring people are informed about scientific issues," was the objective that scientists listed as the highest priority. This result was again confirmed by Rose et al.,[8] in which scientists in their survey of faculty at land-grant universities ranked the same objective the highest. While it is promising that scientists are open to different communication objectives, the deficit approach to communication, where more information and increasing knowledge will lead to a desired goal, continues to be instilled in the community. Facts are always open to interpretation from the audience if they are not put into a context that they can relate to. And for politicized science issues more specifically, the relationship between science knowledge and attitudes is zero. In some cases, it is actually those individuals with the highest levels of science knowledge that indicate the most politicized attitudes.[6]

While these objectives provide guidance on how a scientist may approach communication more strategically, not all scientists view all objectives equally. Some may feel that they do not have the communication skills to achieve an objective. Tactics—or how a communicator chooses what to say or write and the style of communication, help scientists to achieve their communication objectives. The choice of a specific communication tactic helps scientists achieve intended changes from their specific audience, but also how you, as a scientist, may change as well as a result of communication. These tactics focus on several different themes for science communicators: relation, trust, emotion, and organizational. Relational tactics include speaking in a certain way that helps to connect with an audience, or using specific frames or first-person stories to help connect with an audience. Trust related tactics include making sure that your audience feels listened to and a desire to help the community. Emotion related tactics include using either positive or negative charged emotions (e.g. anger or hope) to help connect with an audience. Finally, organizational related tactics involve seek professional communicators for support or allocating part of a grants budget to communication efforts. In a study of American scientists' attitudes toward these different tactics, scientists indicate they are most willing to use relational and trust tactics, with organizational tactics and the use of negative emotions viewed with less support.[33]

Public communication without careful thought about strategy compromises the ability for the scientific community and scientists to reach their communication goals. The emphasis on strategic communication is increasingly critical as the relevance of science in our day-to-day lives, and the pace of technological advances, requires clear and effective communication with various stakeholders. As scientists indicate a willingness to embrace a variety of communication objectives and tactics, it is imperative to understand what combination of objectives and tactics might work best for reaching a specific goal. As scientists continue to understand the importance of not only seeking out ways to improve their communication skills, but understanding how strategy, when applied to these skills, is beneficial, so too is it critical to understand how both scientists and the public think and feel about science. Calls for more communication from the scientific community are not just about ensuring adequate funding for science or how the best available science is used in policy-making, it also is about managing how the public connects to science. In turn, this may shed important contexts for thinking about what are the common underlying ways that both scientists and non-scientists think and feel about science. Any type of universal baseline for how broad segments of the public connect to science, could serve as an important foundation for communication strategies to build from.

4. The brand of science and why it matters

As science communication has adapted frameworks from public relations and strategic communication to identify ways in which scientists can more effectively reach the goals they want to achieve, so to do theories and frameworks from marketing and more specifically, branding. Branding seeks to uncover the emotional, sensory, and cognitive meanings when a person first encounters an object, person, or idea. Various entities carry out extensive research to understand their brand image within the consumer marketplace and use this knowledge to differentiate and position in the minds of consumers.[34] In other words, what needs does a brand meets for a consumer.

Any good marketing consultant will tell you that the first step to a successful marketing campaign is knowing thyself. How are you perceived and why? Within the context of the public communication activities of the scientific community, this goes beyond just looking at the perceived risk or benefits of science, optimism or pessimism of science, or overall levels of support for science. Rather it is about understanding the image of science. One avenue of research that provides a promising direction are the positive and negative associations that the public has with science. A reflection on how science is broadly perceived is a missing link within the process of communication outlined above. As much as defining goals, objectives, and tactics for science communicators helped to orient the strategic focus of science communication, so too will defining the image of science that resonate across various publics, both expert and non-expert.

Attempts to address this question started in 2015 with a series of focus groups with the US public. Participants were asked to discuss what thoughts and feelings come to mind when they think of science.[35] Through these focus groups, responses started to cluster along a distinct set of evoked responses to science, including emotions such as "fear," "hope," joy," "caution," and "boredom." From these insights, survey constructs were developed to gauge how members of the US public think and feel about science, and if any differences emerge when controlling for demographic factors. Across surveys of representative samples in 2019 and 2020, hope—and its various synonyms—is the answer that is brought up most frequently.[15,34] What was most unique about this finding, is that even controlling for levels of education and political ideology, hope still remained the most common choice for nearly 50% of respondents, with the next closest response, "joy," receiving between 13–19%.

Hope is a complex emotion and is not new to science communication research.[36] For decades, psychologists focused on the conceptualization and measurement of hope.[37,38] Hope is defined as the "process of thinking about one's goals, along with the motivation to move toward and the ways to achieve those goals."[38] It is a feeling or expectation for a certain thing to happen, what psychologists refer to as a "pay-off minded" orientation. This finding aligns with the longstanding strategy of the science communication community that effective communication is about connecting science to something that your audience cares about. In other words, science is a utility, in that it takes on meaning when it is connected to a certain outcome. But most unique to this finding is that it illustrates a connection that is common across the public writ large.

As previously mentioned, however, attempts to categorize science as a monolith may be misleading. Science may evoke spectrums of emotions to respondents based on the specific type of science and the application of that science, each which have unique associations with the public. For example, in a series of pilot studies seeking to address this question with samples of the US public, rather than giving respondents a closed-answer response about what word comes to mind when they think about science, respondents were asked *how often* do the following words come to mind when you think about specific scientific fields such—including chemistry, biology, and astronomy (T.P. Newman, unpublished data). While hope is still the emotion that respondents associate most frequently across different scientific fields, some fields—such as chemistry and astronomy—also evoke higher levels of joy, while other fields, such as biology, evoke higher levels of caution. While future research on this topic seeks to unpack what the different attributes of hope are that connect to science, it is also important to recognize how other emotions may be salient and why. Science communication and engagement, therefore, must take the approach of *sciences* communication, as each field of science has its own challenges and opportunities to take into consideration.

It is a quite a powerful finding for scientists seeking to communicate about science that hope is a common ground in terms of the emotional connect that members of the public have to science. For one, it suggests that scientists can use this baseline connection as a way to help frame their communication efforts through understanding how hope can be used as a messaging tactic. Indeed, scientists indicate a strong willingness to use hope to talk about their science, or science more generally, to different audiences.[33] But do scientists think and feel about science the same way as the public? The same question that was tested on public samples was also tested on surveys of scientists. Across three separate surveys of scientists, one in 2018 and 2021 which were samples of scientists at the 62 universities that make of the Association of American Universities (AAU), and the other a survey in 2018 that was a mix of academic, private, and governmental sector scientists who were affiliated with 27 different scientific societies, they were asked the same question about how they think and feel about science.

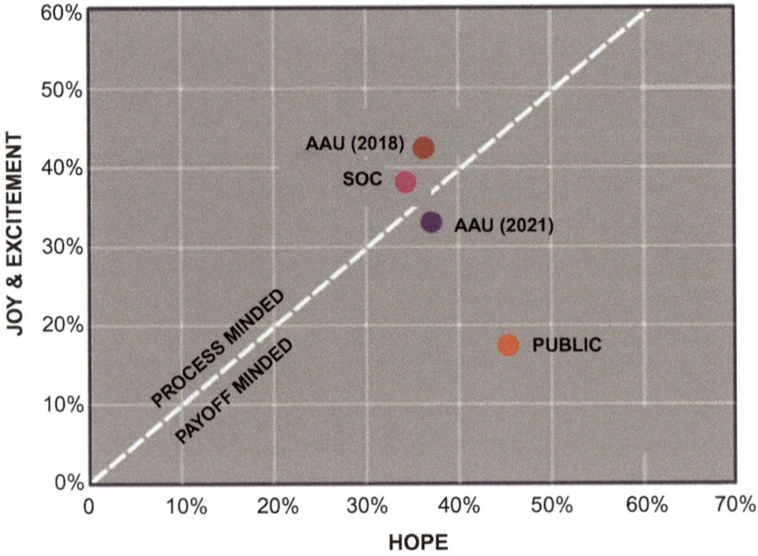

Fig. 1 Process vs payoff minded orientation of scientists and the public. Association of American Universities (AAU): Representative, random sample of U.S. STEM faculty from member universities of the Association of American Universities (AAU): Fall 2018 ($N = 516$); Winter 2021 ($N = 740$). Society (SOC): Sample of U.S. members of 27 professional STEM societies: Fall 2018 ($N = 3619$). Public: Representative, U.S. sample: Fall 2019 ($N = 1525$); Fall 2020 ($N = 1559$).

What the results show is a less consistent pattern: while just under 20% of the public responded "joy," it was much higher for scientists, with nearly 40% answering "joy". Yet, slightly more than two-thirds of scientists also responded hope (Fig. 1).

While hope was overwhelming the emotion most connected to science among the public respondents surveyed, there was more nuance in how scientists answered the question. And indeed, there differences were driven by academic discipline. Scientists in disciplines such as math, physics, and biology—which may be classified as fields that address more basic science research questions, were the most likely to respond joy, and least likely hope. To further unpack these differences, respondents were also segmented by asking them to indicate to what extent their research address basic science research and applied research questions. Again, those who indicate that the majority of their research seeks to address basic research questions were more likely to indicate joy than hope, while those who focused on applied research questions, more likely to indicate hope than joy.

In contrast to the payoff-minded orientation of hope, joy suggests a "process-minded" orientation, where the day-to-day experience of conducting scientific research motivates the emotional response. This is not surprising: Most scientists enjoy the work that they do. Likewise, it gives some support to why increasing wonder and awe of science continues to be a strong goal for scientists in how they approach their communication. This gap between how scientists and non-scientists think and feel about science has several important implications for how one group communicates with the other about the scientific enterprise. Joy and excitement for scientists might shape how they think about science, and indeed may serve as a strong motivation for wanting to communicate with different publics. But joy and excitement about science, while no doubt a way public audiences think and feel about science, is not how the majority of the public connects to science. Rather science is about how it connects to their day-to-day life.

5. Conclusion

Science is political, but it doesn't need to be partisan. The research synthesized in this chapter provides several ways in which researchers focused on science communication and public engagement examine the ways that scientists can take strategic steps to embrace the political nature of science through forms of dialogue and broader stakeholder involvement with different publics. In the current context of "post-normal" science, in which societal implications of science are far reaching, the need for these types of communication activities could not be greater, or more pressing. It may seem ironic in a chapter on the topic of public engagement and communication that increasing the number of scientists who are taking part in communication and engagement activities will not necessarily lead to the desired change in goals that many members of the medical and life science communication may seek to reach. Rather than the quantity of those taking part in communication and engagement activities, it is about the quality. This includes developing the necessary skills for effective communication, but also recognizing the importance of a strategic orientation towards achieving desired goals. Moreover, as the differences in how both scientists and the public connect on an emotional level to science, it is a careful reflection on an important dilemma for all science communication and engagement: should the public come to view science through the experiences of scientists, or should the scientists come to view science through how the public thinks and feels about science. For matters of public communication and engagement, the latter is more important. Scientists need to meet the public where they are in terms of attitudes, values, beliefs, and experiences with science. On a final note, scientists must not take for granted the high-levels of support that the public indicates for the scientific community. Indeed these findings are misleading. It is the application of the science that matters to the public, and once those differences are made clear, we tend to support the science that is connected to outcomes that we view as a priority. Rather as this chapter sought to present, hope is a more stable and potentially fruitful path forward to a common foundation that all of those interested in communicating about the sciences can build from.

Acknowledgments

Todd P. Newman, Ph.D. is an assistant professor in the Department of Life Sciences Communication at the University of Wisconsin-Madison and an affiliate of the Robert F. and Jean E. Holtz Center for Science & Technology Studies and the Nelson Institute for Environmental Studies. His research and teaching focuses on the role of strategic communication within the context of science, technology, and the environment.

References

1. Leshner AI. Public engagement with science. *Science*. 2003;299(5609):977–978.
2. Cicerone RJ. Ensuring integrity in science. *Science*. 2010;5966:624.
3. National Academies of Sciences, Engineering, and Medicine. *Communicating Science Effectively: A Research Agenda*. National Academies Press; 2017.
4. Scheufele DA. Communicating science in social settings. *Proc Natl Acad Sci USA*. 2013;110(Suppl 3):14040–14047.
5. Funtowicz SO, Ravetz JR. Three types of risk assessment and the emergence of post-normal science. In: Krimsky S, Golding D, eds. *Social Theories of Risk*. Westport, CT: Praeger; 1992:251–274.
6. Scheufele DA. Science communication as political communication. *Proc Natl Acad Sci USA*. 2014;111(suppl 4):13585–13592.
7. Besley JC, Dudo A, Yuan S, Lawrence F. Understanding scientists' willingness to engage. *Sci Commun*. 2018;40(5):559–590.
8. Rose KM, Markowitz EM, Brossard D. Scientists' incentives and attitudes toward public communication. *Proc Natl Acad Sci USA*. 2020;117(3):1274–1276.
9. Scheufele DA, Krause NM, Freiling I, Brossard D. What we know about effective public engagement on CRISPR and beyond. *Proc Natl Acad Sci USA*. 2021;118(22).
10. Besley JC, Newman TP, Dudo A, Tiffany LA. Exploring scholars' public engagement goals in Canada and the United States. *Public Underst Sci*. 2020;29(8):855–867.
11. Bennett N, Dudo A, Yuan S, Besley J. Scientists, trainers, and the strategic communication of science. In: Newman TP, ed. *Theory and Best Practices in Science Communication Training*. Routledge; 2019:9–31.
12. National Science Board. *Science and Technology: Public Attitudes and Public Understanding (Chapter 7)*. Science and Engineering Indicators; 2018.
13. Krause NM, Brossard D, Scheufele DA, Xenos MA, Franke K. Trends—Americans' trust in science and scientists. *Public Opin Q*. 2019;83(4):817–836.

14. Krause NM, Scheufele DA, Freiling I, Brossard D. The trust fallacy: Scientists' search for public pathologies is unhealthy, unhelpful, and ultimately unscientific. *Am Sci.* 2021; 109(4):226–232.
15. Newman T, Volpe C, Lindenfeld L, Besley J, Dudo A, Leavey N. *Assessing Scientists Willingness to Engage in Science Communication*; 2019.
16. Newman TP, ed. *Theory and Best Practices in Science Communication Training.* Routledge; 2019.
17. Jensen P. A statistical picture of popularization activities and their evolutions in France. *Public Underst Sci.* 2011;20(1):26–36.
18. Marcinkowski F, Kohring M, Fürst S, Friedrichsmeier A. Organizational influence on scientists' efforts to go public: an empirical investigation. *Sci Commun.* 2014;36(1):56–80.
19. Besley JC, Oh SH, Nisbet M. Predicting scientists' participation in public life. *Public Underst Sci.* 2013;22(8):971–987.
20. Besley JC, Dudo AD, Yuan S, Abi GN. Qualitative interviews with science communication trainers about communication objectives and goals. *Sci Commun.* 2016; 38(3):356–381.
21. Cameron C, Collie MC, Baldwin CD, et al. The development of scientific communication skills: A qualitative study of the perceptions of trainees and their mentors. *Acad Med.* 2013;88(10).
22. Rowland S, Hardy J, Colthorpe K, Pedwell R, Kuchel L. CLIPS (communication learning in practice for scientists): a new online resource leverages assessment to help students and academics improve science communication. *J Microbiol Biol Educ.* 2018;19(1).
23. Rodgers S, Wang Z, Maras MA, et al. Decoding science: development and evaluation of a science communication training program using a triangulated framework. *Sci Commun.* 2018;40(1):3–32.
24. Dudo A. Toward a model of scientists' public communication activity: the case of biomedical researchers. *Sci Commun.* 2013;35(4):476–501.
25. Stylinski C, Storksdieck M, Canzoneri N, Klein E, Johnson A. Impacts of a comprehensive public engagement training and support program on scientists' outreach attitudes and practices. *Int J Sci Educ Part B.* 2018;8(4):340–354.
26. Dudo A, Besley J, Kahlor LA, Koh H, Copple J, Yuan S. Microbiologists' public engagement views and behaviors. *J Microbiol Biol Educ.* 2018;19(1):19.
27. Rubega MA, Burgio KR, MacDonald AA, Oeldorf-Hirsch A, Capers RS, Wyss R. Assessment by audiences shows little effect of science communication training. *Sci Commun.* 2021;43(2):139–169.
28. Copple J, Bennett N, Dudo A, et al. Contribution of training to scientists' public engagement intentions: a test of indirect relationships using parallel multiple mediation. *Sci Commun.* 2020;42(4):508–537.
29. Ecklund EH, James SA, Lincoln AE. How academic biologists and physicists view science outreach. *PLoS One.* 2012;7(5):e36240.
30. Baram-Tsabari A, Lewenstein BV. Science communication training: what are we trying to teach? *Int J Sci Educ Part B.* 2017;7(3):285–300.
31. David YB, Baram-Tsabari A. Evaluating science communication training: Going beyond self-reports. In: Newman TP, ed. *Theory and Best Practices in Science Communication Training.* Routledge; 2019:122–138.
32. Besley JC, Dudo A, Yuan S. Scientists' views about communication objectives. *Public Underst Sci.* 2018;27(6):708–730.
33. Besley JC, Newman TP, Dudo A, Tiffany LA. American Scientists' willingness to use different communication tactics. *Sci Commun.* 2021.
34. Newman BI, Newman TP. *Brand.* Dubuque, IA: Kendall Hunt; 2018.
35. ScienceCounts. *Raising Voices for Science: Exploratory and Benchmarking Survey*; 2015.

36. Nabi RL, Gustafson A, Jensen R. Framing climate change: exploring the role of emotion in generating advocacy behavior. *Sci Commun*. 2018;40(4):442–468.
37. Snyder CR, Lopez SJ, Shorey HS, Rand KL, Feldman DB. Hope theory, measurements, and applications to school psychology. *Sch Psychol Quart*. 2003;18(2):122.
38. Snyder CR. Conceptualizing, measuring, and nurturing hope. *J Couns Dev*. 1995; 73(3):355–360.

Index

Note: Page numbers followed by "*f*" indicate figures and "*t*" indicate tables.

A

Accommodation, 34
Advisory Committee on Immunization Practices (ACIP), 103
Age, 163
  politicization, 181–183
Agencies, 74
  resources/organization, 15
Analytic approach, 114
Antibody tests, 10–11
Anti-intellectualism, 147, 153, 154*f*, 161
Anti-science actions
  agencies, 74
  bias, 72–74
  Biden administration's approach, 74–76
  budget cuts, 70–71
  censorship, 67–69
  misrepresentation, 72–74
  personnel changes, 69–70
  research hindrance, 71
  science restoring, 76–77
  self-censorship, 67–69
Association of American Universities (AAU), 225–226

B

Backlash, 113
Bias, 72–74
Biden administration's approach, 74–76
Biomedical Advanced Research and Development Authority (BARDA), 11–12
Boundary organizations, 35
Budget, 7
  cuts, 70–71

C

Censorship, 67–69
Centers for Disease Control (CDC), 83
Chloroquine, 11–13
Clustered Regularly Interspaced Short Palindromic Repeats (CRISPR) technologies, 206
  age of politicization, 181–183
  communicating science, 172–173
  community-based participatory research (CBPR), 179
  deaf communities, 180
  decision-making, 178–181
  dialogue, 173
  Down syndrome, 180–181
  Environmental Protection Authority (EPA), 181
  NASEM, 178–179
  protective cognition, 183–187
    abstract *vs.* concrete cognition, 184–185
    analogies, 187
    frankenfood framing, 185
    genetically modified organisms (GMO), 185
    involving, 186
    leveraging, 186
    priming different identities, 183–184
    science communication, LIVA model for, 186–187
    visualizing, 187
  public engagement, 173, 181–183
  quality data, lack of, 174–178
  religiosity, 176
  science communications, 177
  shared knowledge production, 178–181
  sociodemographics, 175–176
  speed of advancement, 173–174
  US National Institute for Health (NIH), 173
Coefficient plot, 152*f*
Collaborative cross-disciplinary work, 93–94
Commitments, 200
Communication
  science, 172–173
  training, 217
Communicators, 205

Community-based participatory research (CBPR), 179
Community norms, 139
Comparative effects, 153–155
Complacency, 138
Conceptualization, 47–49
Confusion, 113
Congressional staff, 36–40
Conspiracy
  belief, 148–152
  conspiratorial thinking, 144–147, 160
  predisposition, 160
  theories, 139–141, 144
Contestation, 208
Control agency scientific analyses, 5–7
Control variables, 147–148
COVID-19 pandemic, 208
  conspiracy theory belief, 158–160
  early testing, 9–10
  vaccination
    Centers for Disease Control (CDC), 83
    collaborative cross-disciplinary work, 93–94
    development, 82–83
    immunization cards, 86
    media coverage, 84
    Middle East respiratory syndrome (MERS), 83
    nonpharmaceutical interventions (NPIs), 85
    Operation Warp Speed, 86
    partisan differences in response, 85
    Pfizer/BioNTech, 83
    politicians statements/visual cues, 83
    practitioners of science, 92–94
    severe acute respiratory syndrome (SARS), 83
    social science research, 90–92
    The Trump administration, 85–86
    U.S. Food and Drug Administration, 82, 86
    vaccine acceptance/resistance variability, 87
    vaccine resistant, 85
    variants and continued politicization, 89–90
    World Health Organization (WHO), 89

D
Deaf communities, 180
Decision-making, 178–181
Dependent variable, 144
Descriptive statistics, 145–146*t*
Disinterestedness, 31
Domain specific policy action, 112–113
Down syndrome, 180–181

E
Education, 164
Emergency use authorizations (EUAs), 10–11
Environmental Protection Authority (EPA), 181
External efficacy, 161

F
Federal government, science and politics in, 8–14
  antibody tests, 10–11
  Biomedical Advanced Research and Development Authority (BARDA), 11–12
  chloroquine, 11–13
  COVID-19, early testing for, 9–10
  emergency use authorizations (EUAs), 10–11
  hydroxychloroquine, 11–13
  premature reopening, 13–14
  risks, 9
Flu panic, 84–85
Framing, 205
Freedom of Information Act (FOIA), 6

G
Gender, 164
Genetically modified organisms (GMOs), 207

H
HPV vaccine, 111
Hydroxychloroquine, 11–13
Hyperpolitical settings, 35–36
Hypotheses, 143

I

Ideal institutional design, 18–19
Immunization cards, 86
Income, 164
Independent variables, 144–148
Inoculation, 115–118
  backlash responses, 119–121
  confusion, 119–121
  domain specific policy support, 118–119
  negative emotional reactions, 116*f*
  potential carryover effects, 122–125, 124*f*
  T-test, 124–125*t*
Institutionalization, 4–8, 14–17, 31
Integrated science assessment (ISA), 20–21, 21*f*
Internal efficacy, 161

J

Journalists, 52–54

K

Knowledge-broking strategies, 55
Knowledge societies, 209

L

Learned helplessness, 162
Local advocate summing up, 111

M

Max Weber, 31
Media coverage politicization, 49–56, 84
  Advisory Committee on Immunization Practices (ACIP), 103
  analytic approach, 114
  backlash, 113
  concerns, 104–105
  confusion, 113
  consequences, 104–105
  data and methods, 109–114
  dimensions, 105–106
  domain specific policy action, 112–113
  headlines and inoculation treatments for, 110*t*
  health recommendations, 103–104
  HPV vaccine, 111
  inoculation, 115–118
    backlash responses, 119–121
    confusion, 119–121
    domain specific policy support, 118–119
    negative emotional reactions, 116*f*
    potential carryover effects of, 122–125, 124*f*
    T-test, 124–125*t*
  local advocate summing up, 111
  measures, 112–114
  media ecosystem, 102–103
  messaging strategies to, 106–107
  negative emotional responses, 112, 115–118
  policy attitudes, 102
  potential carryover effects, 105, 114
  Preventive Services Task Force, 103
  public attitudes, 102
  science of guidelines, 103–104
  study design, 109–111
  study objectives, 107–108
Media ecosystem, 102–103
Middle East respiratory syndrome (MERS), 83
Misinformation, 139–141
Misrepresentation, 72–74
Moral coherence seeking, 201
Moral conviction
  biased assimilation, 199
  challenges, 208–210
  commitments, 200
  communicators, 205
  contestation, 208
  COVID-19 pandemic, 208
  CRISPR, 206
  definition, 196–198
  disagreements, 198
  framing, 205
  genetically modified organisms (GMOs), 207
  issue attitude, 201–202, 204*f*
  knowledge societies, 209
  moral coherence seeking, 201
  Pilot survey, 203*t*
  politicization of science, 198–202
  pre-existing attitudes, 200–201
  science issues, public opinion on, 202–204
  science of science, 209–210

Moral conviction (*Continued*)
scientific activity, 195
self-interested reasons, 208
social science research, 195–196
strategic communication, 205–208
war on science, 195

N

NASEM, 178–179
National ambient air quality standards (NAAQS), 19–20, 23
National Science Foundation, 32–33
Non-democratic governments, 47
Non-epistemic values, 32
Nonpharmaceutical interventions (NPIs), 85
Normal flu, 84–85

O

Office of Science and Technology Policy (OSTP), 34
Office of Technology Assessment (OTA), 35
Operation Warp Speed, 86
Organization for Economic Cooperation and Development (OECD), 4

P

Participatory turn, 34–35
Partisan differences, 85
Partisanship, 160, 216–217
Pay-off minded orientation, 224–225
Personnel changes, 69–70
Pfizer/BioNTech, 83
Philosophical recognition, 32–33
Pilot survey, 203*t*
Policy attitudes, 102
Policymaking process, 46
Political identity, 147
Political ideology, 160–161
Political interest, 163
Political knowledge, 162–163
Political weapon, 84–85
Politicians statements/visual cues, 83
Politicization of science
conceptualization, 47–49
definition of, 47–49
journalists, 52–54
knowledge-broking strategies, 55
media, 49–56
non-democratic governments, 47
policymaking process, 46
political actors, 50–52
politicization processes, 52
politics, 49–56
practical implications, 57–59
science, 54–56
scientific actors, 54–56
scientific evidence, 46
triangular relationship, 49–56
Politicization of science, US
budget, 7
control agency scientific analyses, 5–7
federal government, science and politics in, 8–14
antibody tests, 10–11
Biomedical Advanced Research and Development Authority (BARDA), 11–12
chloroquine, 11–13
COVID-19, early testing for, 9–10
emergency use authorizations (EUAs), 10–11
hydroxychloroquine, 11–13
premature reopening, 13–14
risks, 9
Freedom of Information Act (FOIA), 6
ideal institutional design, 18–19
institutional design, 4–8, 14–17
integrated science assessment (ISA), 20–21, 21*f*
manipulating
agency resources/organization, 15
pursuit of an agenda, 14–15
science advisory boards, 15–16
multiple pathways, 14–17
national ambient air quality standards (NAAQS), 19–20, 23
public deliberations, 16–17
reformed design, 19–23
risk/exposure assessment, 22
science-intensive decision-making, 8
scientific independence, 4
staffing, 7
well-endowed stakeholders, 16–17
Politicization processes, 52

Post-normal science, 215–216
Practitioners of science, 92–94
Pre-existing attitudes, 200–201
Premature reopening, 13–14
Preventive Services Task Force, 103
Protective cognition, 183–187
abstract *vs.* concrete cognition, 184–185
analogies, 187
frankenfood framing, 185
genetically modified organisms (GMO), 185
involving, 186
leveraging, 186
priming different identities, 183–184
science communication, LIVA model for, 186–187
visualizing, 187
Protective silos
disinterestedness, 31
institutionalization, 31
Max Weber, 31
non-epistemic values, 32
objectivity, 31–32
Salk Institute, 31
science policy, 32–35
social construction of scientific facts, 31
Public attitudes, 102
Public communication, scientists attitudes and behavior, 217–221
Public deliberations, 16–17
Public engagement, 181–183, 218
Pursuit of an agenda, 14–15

Q

Quality data, lack of, 174–178

R

Race, 163
Reformed design, 19–23
Re-imagining scientists engagement, 40–41
Religiosity, 176
Research hindrance, 71
Risk/exposure assessment, 22

S

Salk Institute, 31
Satisfaction, 162
Science, 32–35, 54–56, 223–227
accommodation, 34
advisory boards, 15–16
boundary organizations, 35
communications, 177
congressional staff, 36–40
guidelines, 103–104
hyperpolitical settings, 35–36
issues, public opinion on, 202–204
legislatures, 35–36
Mode 1, 34
Mode 2, 34
Mode 3, 34
National Science Foundation, 32–33
Office of Science and Technology Policy (OSTP), 34
Office of Technology Assessment (OTA), 35
participatory turn, 34–35
re-imagining scientists engagement, 40–41
science and technology (S&T) policy, 33
World War II, 32–33
Science and technology (S&T) policy, 33
Science-intensive decision-making, 8
Science of science, 209–210
Science restoring, 76–77
Scientific activity, 195
Scientific actors, 54–56
Scientific evidence, 46
Scientific independence, 4
Select Subcommittee on the Coronavirus Crisis, 84
Self-censorship, 67–69
Self-interested reasons, 208
Severe acute respiratory syndrome (SARS), 83
Shared knowledge production, 178–181
Silencing Science Tracker, 66–67
Social construction of scientific facts, 31
Social media messages, 85
Social science research, 90–92, 195–196
Sociodemographics, 175–176
Speed of advancement, 173–174
Staffing, 7
Staff members, 37–39
STEM education, 217

Strategic science communication, 205–208
communication training, 217
connecting goals, 221–223
objectives, 221–223
partisanship, 216–217
post-normal science, 215–216
public communication, scientists attitudes and behavior, 217–221
public engagement, 218
science, 223–227
STEM education, 217
tactics, 221–223
trust, 216–217
willingness prediction to engage, 218–221

T

Tactics, 221–223
The Trump administration, 85–86
Trust, 161, 216–217

U

U.S. Congress, 29–30, 33, 36–37, 41
U.S. Food and Drug Administration, 82, 86
US National Institute for Health (NIH), 173

V

Vaccine acceptance/resistance variability, 87
Vaccine hesitancy, 158
anti-intellectualism, 147, 153, 154*f*
coefficient plot, 152*f*
community norms, 139
comparative effects, 153–155
complacency, 138
conspiracy belief on, 148–152
conspiracy theories, 139–141
conspiratorial thinking, 144–147
control variables, 147–148
data, 143
dependent variable, 144
descriptive statistics, 145–146*t*
expected correlates of, 141–142
hypotheses, 143
independent variables, 144–148
measures, 144
misinformation, 139–141
political identity, 147
uptake, 137–139
vaccine-resistant strains, 137
World Health Organization (WHO), 136–137
Vaccine resistant, 85, 137
Vaccine support
anti-intellectualism, 153
conspiracy belief on, 148–152
dependent variable, 144
political identity, 153
Variants and continued politicization, 89–90

W

War on science, 195
Well-endowed stakeholders, 16–17
Willingness prediction to engage, 218–221
World Health Organization (WHO), 89, 136–137
World War II, 32–33

FSC
www.fsc.org
MIX
Papier | Fördert
gute Waldnutzung
FSC® C083411

Kolibri 360 GmbH
Ferdinand-Jühlke-Straße 7
99095 Erfurt, Deutschland
produktsicherheit@kolibri360.de